ALLE · ZEIT · WACH
1842

Walter Brendel

K. Messmer M. Stein (Eds.)

Pathways in Applied Immunology

In Memoriam Walter Brendel

Springer-Verlag
Berlin Heidelberg New York
London Paris Tokyo
Hong Kong Barcelona
Budapest

Prof. Dr. Konrad Messmer
Mechthild Stein

Institute for Surgical Research,
Klinikum Grosshadern, Ludwig-Maximilians-University,
Marchioninistr. 15, W-8000 Munich 70, FRG

With 27 Figures

ISBN-13:978-3-642-76608-4 e-ISBN-13:978-3-642-76606-0
DOI: 10.1007/978-3-642-76606-0

Library of Congress Cataloging-in-Publication Data
Pathways in applied immunology: in memoriam Walter Brendel/K. Messmer, M. Stein (eds.). p. cm.
Revised papers of the 20th Round Table Symposium on Applied Immunology, held in Kitzbühel, Austria, in 1989. "Selected publications by Walter Brendel": p. Includes index.
ISBN-13:978-3-642-76608-4

1. Clinical immunology—Congresses. 2. Brendel, Walter—Congresses. I. Brendel, Walter. II. Messmer, K. III. Stein, M. (Mechthild), 1923–. IV. Round Table Symposium on Applied Immunology (20th: 1989: Kitzbühel, Austria)
RC581.2.P38 1991 616.07'9—dc20 91-17390

Softcover reprint of the hardcover 1st edition 1991

Photograph on page II by Hans Deumling, Schleissheimer Str. 68, W-8000 Munich 80, FRG

Typesetting: Best-set Typesetter Ltd., Hong Kong
27/3130-543210 – Printed on acid-free paper

Preface

This book is dedicated to the memory of Walter Brendel, late Professor of Experimental Surgery and Chairman of the Institute for Surgical Research at the University of Munich, Germany. For 20 years Walter Brendel organized the renowned Round Table Symposium on Applied Immunology, first in Kitzbühel and later in Axams, Austria. On the occasion of the 20th symposium in January 1989 he gathered together a number of scientists who have been leaders in the field of transplantation immunology and clinical transplantation for the past two decades. All of them had participated at previous meetings, some on a regular basis.

Many of the new discoveries in applied immunology and transplantation medicine were first presented and vividly discussed at the Round Table Symposia. The annual Kitzbühel/Axams meetings became well-known and invitations much sought after, not only for this reason but also because of the uniquely intimate atmosphere that promoted the free exchange of research findings and theoretical cut and thrust.

From the onset, in the late 1960s, Walter Brendel and his coworkers provided an ambience of scientific endeavour combined with sporting activity on the ski slopes of the Tyrol. A great esprit de corps developed among the participants, based not only on sharing the pleasure of new discoveries and scientific achievements but also on experiencing difficult situations in the mountains and coping with aches and pains caused by less than masterly skiing. The kind of meeting initiated by Walter Brendel is now very popular and has been adopted by other groups and associations.

Walter Brendel was deeply concerned about the inability of many scientists to communicate with colleagues in other disciplines, and he went out of his way to bring together scientists as far apart in their interests as basic physiology, surgical pathophysiology, and immunology and molecular biology as applied to transplantation medicine. It is this multidisciplinary approach to biological and medical problems that has given these meetings their unique character. Although attendance was entirely by invitation and numbers were restricted to 35–40 participants, the ideas and data presented were made available to the scientific community as abstracts published in the journal *European Surgical Research*, edited by Walter Brendel jointly with Konrad Messmer. As a special event the presentations and discussions of the 20th symposium were tape-recorded and edited by the authors and editors for full publication.

At the 1989 meeting in Kitzbühel Walter Brendel was already suffering from a severe and incurable illness to which he succumbed in August of that year. It was then decided by the participants that the presentations at the 20th symposium should be published in the form of a book dedicated to his memory. The participants received typewritten versions of their presentation including their remarks addressing Walter Brendel himself, thus giving their contributions an unusually personal as well as nostalgic flavour. The editors of this book were very pleased to receive revised versions and these are gathered together in this volume. The contributions cover a wide range and give an overview of the development of specific fields in applied immunology. Lady Jean Medawar contributes recollections of her late husband, the Nobel Laureate Sir Peter Medawar. Sir Peter participated in the symposia of the first decade and exerted a significant influence on them. Professor Leslie Brent reflects upon the ups and downs of a scientist.

Very typically for Walter Brendel, he felt that even a group of scientists fully dedicated to immunology should not listen to and exclusively discuss immunology during a 3-day meeting. He therefore always provided a carefully selected "Relaxing from Immunology" lecture, given by outstanding individuals on topics as diverse as "Bats, Truffle Pigs and Sex Hormones", "Allergy", "X-Ray Crystallography of Proteins", "Thrombosis Prophylaxis", "High Altitude Research" and "Brain Edema". For the 20th meeting he had invited a long-time friend from his earlier career as physiologist, Professor Jürgen Aschoff, former director of the Max Planck Institute for Behavioral Physiology, Seewiesen and one of the founders of chronobiology. Professor Aschoff's lecture was received with great enthusiasm by the participants; he addressed the question "Why should we rise with the lark?", and the answer to this question, which seems particularly relevant for scientists, is to be found in the last chapter of this book.

As former collaborators of Walter Brendel and as editors we would like to thank the authors who have contributed to this volume and the staff of Springer-Verlag who made publication possible.

It is our hope that this volume will give the scientific community an insight into the spirit and scientific life of Walter Brendel. For this reason his curriculum vitae and a list of his most important publications encompassing the wide gamut of his research interests have been included. We hope that the book will help the reader to understand why Walter Brendel was respected so highly, not only by his pupils and collaborators, but also by a great number of contemporary scientists from many countries.

Munich, August 1991

Konrad Messmer
Mechthild Stein

Contents

Contributors

J. Aschoff, Prof. Dr.
Jacobistrasse 29, W-7800 Freiburg/Br., FRG

F. H. Bach, Prof., M.D.
Immunology Research Center, Department of Laboratory Medicine and Pathology, University of Minnesota, Twin Cities Medical School, Box 724, 420 Delaware Street, S.E., Minneapolis, MN 55455, USA

J. R. Batchelor, Prof., M.D.
Department of Immunology, Royal Postgraduate Medical School, Hammersmith Hospital, Du Cane Road, London W12 0NN, United Kingdom

D. W. van Bekkum, Prof., M.D.
Institute for Applied Radiobiology and Immunology TNO, Postbus 5815, NL-2280 HV Rijswijk, The Netherlands

L. Brent, Prof., Ph.D.
Transplant Unit, 4th Floor Clarence Wing, St. Mary's Hospital, London W2 1NY, United Kingdom

Sir R. Calne, Prof., M.B.
University of Cambridge, Clinical School, Department of Surgery, Level 9, Addenbrooke's Hospital, Hills Road, Cambridge CB2 2QQ, United Kingdom

C. H. Chester, M.D.
Immunology Branch, National Cancer Institute, National Institutes of Health, 9000 Rockville Pike, Bethesda, MD 20892, USA

I. Funke, Dr.
Institut für Immunologie, Technische Universität München,
Goethestrasse 31, W-8000 München 2, FRG

F. Hammersen, Prof. Dr.†
formerly: Anatomisches Institut,
Technische Universität München, W-8000 München 40, FRG

K. Heeg, Prof. Dr.
Institut für Medizinische Mikrobiologie und Hygiene,
Technische Universität, Klinikum rechts der Isar,
Trogerstrasse 9, W-8000 München 80, FRG

Jean Lady Medawar
25 Downshire Hill, London NW3 1NT, United Kingdom

M. D. Menger, Dr.
Institut für Chirurgische Forschung,
Klinikum Grosshadern, Ludwig-Maximilians-Universität,
Marchioninistrasse 15, W-8000 München 70, FRG

K. Messmer, Prof. Dr.
Institut für Chirurgische Forschung,
Ludwig-Maximilians-Universität, Klinikum Grosshadern,
Marchioninistrasse 15, W-8000 München 70, FRG

N. A. Mitchison, Prof.
Deutsches Rheuma-Forschungszentrum Berlin,
Am kleinen Wannsee 5, 1000 Berlin 39, FRG

A. P. Monaco, Prof., M.D.
Department of Surgery, Harvard Medical School and
New England Deaconess Hospital,
185 Pilgrim Road, Boston, MA 02215, USA

P. J. Morris, Prof., M.B., Ph.D.
Nuffield Department of Surgery, University of Oxford,
John Radcliffe Hospital, Headington,
Oxford OX3 9DU, United Kingdom

J. A. Myburgh, Prof., M.D.
Department of Surgery, University of the Witwatersrand,
Johannesburg Medical School, York Road,
Parktown 2193, South Africa

K. Pantel, Dr.
Institut für Immunologie, Technische Universität München,
Goethestrasse 31, W-8000 München 2, FRG

K. Pfeffer, Dr.
Institut für Medizinische Mikrobiologie und Hygiene,
Technische Universität, Klinikum rechts der Isar,
Trogerstrasse 9, W-8000 München 80, FRG

G. Riethmüller, Prof. Dr.
Institut für Immunologie, Technische Universität München,
Goethestrasse 31, W-8000 München 2, FRG

J. J. van Rood, Prof., M.D., Ph.D.
Department of Immunohematology and Blood Bank, Building 1,
E-3Q, University Hospital, P.O. Box 9600,
NL-2300 RC Leiden, The Netherlands

D. H. Sachs, Prof., M.D.
Massachusetts General Hospital, MGH East, Building 149,
13th Street, Charlestown, MA 02129, USA

G. Schlimok, Dr. med.
Institut für Immunologie, Technische Universität München,
Goethestrasse 31, W-8000 München 2, FRG

Elisabeth Simpson, M.A., Vet. M.B.
Clinical Research Centre, Transplantation Biology Section,
Watford Road, Harrow, Middlesex HA1 3UJ, United Kingdom

M. Sykes, M.D.
Immunology Branch, National Cancer Institute,
National Institutes of Health,
9000 Rockville Pike, Bethesda, MD 20892, USA

H. Wagner, Prof. Dr.
Institut für Medizinische Mikrobiologie und Hygiene,
Technische Universität, Klinikum rechts der Isar,
Trogerstrasse 9, W-8000 München 80, FRG

Professor Dr. med. Dr. med. h. c. Walter Brendel 1922–1989

Walter Brendel was born as second son to a company director on November 6, 1922 in Karlsruhe. He attended prep school for 4 years and the Goethe Gymnasium in Karlsruhe for 8 years, passing his school-leaving examination in 1941. Soon after entering the Goethe Gymnasium he became a member of the German Boy Scouts. From 1941 to 1944 he served in the 98th Alpine Regiment of the German Army in Russia, where he was severely injured in 1942. Due to his injuries he was released from the front, which gave him the opportunity to start studying medicine in summer 1942 at the University of Heidelberg. He passed his preliminary examination in August 1944, his finals in summer 1948. From 1948 to 1951 he completed his internship in surgery, ophthalmology and internal medicine. He prepared his medical thesis, entitled "Postoperative Parotitis", under the direction of K.H. Bauer, Professor of Surgery at the University of Heidelberg. The thesis was accepted by the Medical Faculty of the University at Heidelberg in 1949.

In March 1951 Walter Brendel became an assistant at the William Kerckhoff Institute for Cardiac Research in Bad Nauheim and at the Institute for Physiology of the University of Giessen under the direction of Professor Hans Schäfer. In that year he became involved in the electrophysiology of the heart. Under the guidance of his principle teacher Professor Rudolf Thauer, Walter Brendel published a series of papers on monitoring of circulatory parameters of the unanesthetized dog, on the influence of anesthesia on the cardiovascular system and metabolism, and on peripheral circulation and temperature regulation. In 1952 he married Jutta, née Weyl. Starting in 1954 he explored the physiology of deep hypothermia; he served as physician to the joint German-Austrian-American expedition to the Himalayas in 1955. After his return he concentrated on the physiology of temperature regulation, in particular on its cerebral control, and became involved in the pathophysiology of deep hypothermia under extracorporeal circulation. In 1959 he became Privatdozent for Physiology at the University of

Giessen. Shortly thereafter he established scientific contacts with cardiovascular surgeons, who were at that time employing extracorporeal circulation in open heart surgery. He met Professor Rudolf Zenker, Chairman of the Department of Surgery of the University of Munich, who proposed that he establish a laboratory for experimental surgery at the University of Munich with the aim of promoting research in pathophysiology and surgery at this institution. In fall 1961 Brendel spent a sabbatical in the United States, including visits to and lectures at the Mayo Clinic, Stanford University, and Amory University in Atlanta.

At the beginning of 1962 Walter Brendel and his family moved from Bad Nauheim to Munich, where he started the Laboratory for Experimental Surgery in the Department of Surgery, the University of Munich Medical School. In accepting this challenge Walter Brendel turned his attention from general physiology to surgical pathophysiology with its myriad fascinating aspects, realizing at an early stage that experimental surgery or surgical research should by no means be limited to the improvement of surgical techniques and decisions, but should involve the implementation of new surgical concepts backed by scientific exploration with emphasis on clinical application. With his first coworkers, Konrad Messmer and Jürgen Reulen, he initiated research on death due to cold, circulation, shock, and wound healing. As early as 1963, with his characteristic pragmatic approach to scientific problems, he engaged his team on research into transplantation, in particular organ transplantation, which at that time seemed to him a more successful approach to coping with organ failure than the use of artificial organs. Together with pupils and clinical collaborators he made important contributions to the immunosuppressive treatment of transplant recipients. From the Laboratory for Experimental Surgery originated, in cooperation with Rudolf Pichlmayr, the Munich antilymphocyte serum and antilymphocyte globulin, which were used in clinical transplantation, in particular by Christian Barnard in the first heart transplants in Cape Town.

In 1968 Dr. Brendel founded – and headed until 1982 – the program "Restitution and Substitution of Internal Organs", sponsored by the *Deutsche Forschungsgemeinschaft*. Within the framework of this program the basis for experimental and clinical transplantation immunology and organ transplantation was initiated and firmly established at the University of Munich, which has evolved today into one of the most active and successful transplant centers in Germany.

In 1966 Walter Brendel received the offer of a Professorship of Clinical Physiology at the Free University of Berlin, but he decided to remain in Munich after the creation for him of an extraordinary Professorship for Experimental Surgery in 1967. After he had turned down the offer of a Professorship for Experimental Surgery at the University of Bonn, the University of Munich endowed Professor Brendel with the first Chair of Experimental Surgery in 1969.

Professor Brendel's Institute for Surgical Research in Munich became internationally recognized for its wide range of research, including programs on surgical pathophysiology, microcirculation, shock, blood replacement and hemodilution, brain edema, secondary brain injury, gastric physiology, wound healing, transplantation cytology, hypothermia, and high altitude medicine. In 1968 Walter Brendel was elected President of the European Society for Experimental Surgery. He became one of the founders of the journal *European Surgical Research* and was its sole Editor in Chief during the first years, later sharing this office with Konrad Messmer from 1978 to 1989. In 1977, in recognition of his pioneering work in organ transplantation the University of Sao Paulo, Brazil, awarded Professor Brendel the title Doctor honoris causa; in 1979 the German Society for Surgery acknowledged his extensive contributions to surgery by bestowing upon him its Jubilee Award; and in 1984 he received the *Bundesverdienstkreuz* (a German national award in recognition of outstanding services).

With his gift for innovative scientific approaches coupled with untiring enthusiasm, in 1975 Walter Brendel tackled the problem of utilizing shock waves in clinical medicine in cooperation with his pupil Christian Chaussy and research fellows from the Department of Urology. Initially this new approach to the very old surgical problem of kidney stones was met with widespread scepticism. Nevertheless, after shock wave lithotripsy in urology became routine Walter Brendel laid the ground for the treatment of biliary stones with shock waves in cooperation with his pupil Michael Delius and his clinical partners Professors Paumgartner and Sauerbruch. Today these two procedures represent a shining example of the importance of surgical research, particularly the introduction of new nonsurgical modalities for treatment of surgical diseases. For his innovative contributions to the application of shock waves in medicine Walter Brendel was honoured in 1985 with the Körber Award for the Advancement of European Science, and he received a significant grant for the exploration of the use of shock waves in the treatment of solid tumors.

In his scientific career Walter Brendel inspired and educated researchers and scholars from all parts of the world. Over the years his Institute for Surgical Research has provided tuition for a great number of research fellows from all medical disciplines from the University of Munich and many other institutions in Germany, Europe, and further afield. Walter Brendel's significant cortributions both to the field of surgical research and to clinical surgery in Germany are reflected by the great number of important positions held by his students and former collaborators throughout the country. Through his own scientific achievements, the establishment of his School of Experimental Surgery, and, in particular, through the long list of scholars and colleagues who received their training in basic research under his guidance, Walter Brendel will be present in the field of the surgical sciences for generations to come.

In 1987 Professor Brendel became seriously ill. He recovered sufficiently to assemble together his closest friends in the field of transplantation medicine and immunology in January 1989 for the 20th Round Table Symposium on Applied Immunology in Kitzbühel, Austria, which he had organized and presided over since 1969, but then succumbed to his illness and died on August 29, 1989 at the Grosshadern University Hospital in Munich.

Selected Publications by Walter Brendel

Brendel W, Raule W, Trautwein W (1950) Die Leitungsgeschwindigkeit und Erregungsausbreitung in den Vorhöfen des Hundes. Pflügers Arch Ges Physiol 253:106

Brendel W, Gladewitz H, Hildebrandt F, Trautwein W (1951) Elektrophysiologische Untersuchungen am Herz-Lungen-Präparat nach Starling. Cardiologia 18:345

Brendel W, Koppermann E (1953) Zur Methode der fortlaufenden unblutigen Blutdruckregistrierung an der Carotisschlinge. Pflügers Arch Ges Physiol 258:121

Brendel W, Koppermann E, Thauer R (1954) Der Kreislauf in Narkose (Seine Beziehung zum respiratorischen Stoffwechsel). Pflügers Arch Ges Physiol 259:357

Brendel W, Koppermann E, Thauer R (1954) Der respiratorische Stoffwechsel in Narkose (Ein Beitrag zur Frage des Minimalumsatzes). Pflügers Arch Ges Physiol 259:177

Brendel W, Koppermann E (1954) Über die Beziehungen zwischen Kreislauf und Stoffwechsel in verschiedenen Narkosestadien. Ber Ges Physiol Exper Pharmakol 162:330

Brendel W (1955) Frankfurter Himalaya-Expedition 1955. Mitteilg Max Planck Ges 5:256

Brendel W (1956) Anpassung von Atmung, Hämoglobin, Körpertemperatur und Kreislauf bei langfristigem Aufenthalt in grossen Höhen (Himalaya). Pflügers Arch Ges Physiol 263:227

Brendel W (1956) Der Gesundheitszustand der Bergvölker im Karakorum. Umschau 16:494

Brendel W, Albers C, Usinger W (1957) Kreislaufregulation in tiefer Hypothermie. Pflügers Arch Ges Physiol 266:45

Brendel W, Usinger W (1959) Die Bedeutung der Gehirntemperatur für die Kältegegenregulation. Pflügers Arch Ges Physiol 270:73

Brendel W (1960) Die Bedeutung der Hirntemperatur für die Kältegegenregulation. I. Der Einfluss der Hirntemperatur auf den respiratorischen Stoffwechsel des Hundes in thermoindifferenter Umgebung. Pflügers Arch Ges Physiol 270:607

Brendel W (1960) Die Bedeutung der Hirntemperatur für die Kältegegenregulation. II. Der Einfluss der Hirntemperatur auf den respiratorischen Stoffwechsel des Hundes unter Kältebelastung. Pflügers Arch Ges Physiol 270:628

Brendel W (1960) Die Bedeutung der Hirntemperatur für die Kältegegenregulation. III. Der Einfluss der Hirntemperatur auf den Kreislauf des Hundes. Pflügers Arch Ges Physiol 270:648

Brendel W, Usinger W (1961) Die Bedeutung der Hirntemperatur für die Auslösung des Kältezitterns. Ein Beitrag zur Frage der cerebralen Kältereception. Pflügers Arch Ges Physiol 274:77

Brendel W (1962) The effect of isolated brain cooling on the cardio-vascular system. Ann Chir Thorac Cardiovasc 1:1

Thauer R, Brendel W (1962) Hypothermie. Prog Surg 2:37

Brendel W, Hallwachs O, Usinger W (1962) Sauerstoffverbrauch und Hämodynamik bei kombinierter Anwendung von Herz-Lungenmaschine und tiefer Hypothermie bis 10°C. Thoraxchir 9:607

Enzenbach R, Brendel W, Messmer K (1963) Möglichkeiten und Grenzen der selektiven Hirnkühlung. Langenbecks Arch Klin Chir 302:136

Brendel W, Enzenbach R, Messmer K (1963) Probleme der isolierten Hirnkühlung und tiefen Gesamtkörperkühlung. Langenbecks Arch Klin Chir 304:739

Brendel W, Baethmann A, Reulen HJ (1965) Die Deutung der Harnstoffwirkung beim Hirnödem. Langenbecks Arch Klin Chir 313:970

Brendel W, Reulen HJ, Messmer K (1965) Die Kälteschwellung des Gehirns und die Begrenzung der Überlebenszeit in Hypothermie. Klin Wochenschr 43:515

Brendel W, Reulen HJ, Aigner P (1965) Die Kältetoleranz des Winterschläfers. Naturwissenschaften 52:501

Reulen HJ, Aigner P, Brendel W, Messmer K (1966) Elektrolytveränderungen in tiefer Hypothermie. I. Die Wirkung akuter Auskühlung bis 0°C und Wiedererwärmung. Pflügers Arch Physiol 288:197

Brendel W, Müller C, Reulen HJ, Messmer K (1966) Elektrolytveränderungen in tiefer Hypothermie. II. Beziehungen zur klinischen und biologischen Überlebenszeit. Pflügers Arch Physiol 288:220

Messmer K, Brendel W, Reulen HJ, Nordmann KJ (1966) Elektrolytveränderungen in tiefer Hypothermie. III. Beziehungen zur biologischen Überlebenszeit bei künstlichem Kreislauf. Pflügers Arch Physiol 288: 240

Brendel W, Reulen HJ, Aigner P, Messmer K (1966) Elektrolytveränderungen in tiefer Hypothermie. IV. Die Kälteschwellung des Gehirns beim Winterschläfer. Pflügers Arch Physiol 292:83

Pichlmaier H, Brendel W (1966) Experimentelle Gesichtspunkte zur Organtransplantation. Klin Wochenschr 44:57

Reulen HJ, Steude U, Brendel W (1966) Die Blutdruckabhängigkeit des zerebralen Energiestoffwechsels bei extrakorporalem Kreislauf, Thoraxchir Vask Chir 14:413

Rinecker H, Brendel W (1966) Erste Ergebnisse fortlaufender Messungen der Darmperistaltik in vivo am wachen oder narkotisierten Hund. Pflügers Arch Physiol 289:43

Messmer K, Brendel W, Devens K, Reulen HJ, Kraemer M (1966) Druckabhängigkeit der Leberdurchblutung. Pflügers Arch Physiol 289:75

Mendler N, Brendel W, Reulen, HJ (1967) Die Kälteschwellung des Gehirns bei hypothermer (2–4°C) Perfusion der Ratte mit einer Herz-Lungen-Maschine. Pflügers Arch Physiol 297:15

Pichlmayr R, Brendel W, Beck G, Schmittdiel E, Tidow G, Pichlmayr I (1967) Herstellung heterologer Immunseren gegen Lymphocyten des Hundes. Klin Wochenschr 45:199

Pichlmayr R, Brendel W, Zenker R (1967) Production and effect of heterologous anti-canine lymphocyte serum. Surgery 61:774

Pichlmayr R, Brendel W, Tidow G (1967) Eigenschaften heterologer Immunseren gegen Hundelymphocyten. Zschr Exper Med 143:299

Pichlmayr R, Brendel W, Wonigeit K (1967) Blutbildveränderungen nach Gabe eines heterologen Antihundelymphocytenserums. Z Exp Med 143:305

Pichlmayr R, Brendel W, Pichlmayr I (1967) Die immusuppressive Wirkung eines heterologen Antihundelymphocytenserums. Z Exp Med 143:313

Mikaeloff P, Pichlmayr R, Rassat JP, Messmer K, Bomel J, Tidow G, Etiennemartin M, Falconnet J, Descotes J, Brendel W (1967) Homotransplantation orthotopique de foie chez le chien: traitement immunodepresseur par serum antilymphocyte. Press Med 75:1967

Messmer K, Brendel W, Holper K (9167) Überleben von Hunden bei akuter Verminderung der O_2-Transportkapazität auf 2,8 g% Hämoglobin. Pflügers Arch Physiol 297:R 48

Sanguinetti F, Messmer K, Devens K, Brendel W (1967) Estudio experimental sobre la circulacion hepatica en el "shock" hemorrhagico y en la hipotension por ganglioplejicos. Rev Argent Cirurg 13:31

Mikaeloff P, Pichlmayr R, Rassat JP, Messmer K, Bomel J, Tidow G, Etiennemartin M, Malluret PJ, Belleville P, Tronchon J, Descotes J, Brendel W (1967) Orthotopic homotransplantation of the liver of the dog – immunodepressive treatment by anti-lymphocyte serum. Excerpta Medica Internat Congr Series 152:38

Zenker R, Brendel W, Pichlmaier H, Pichlmayr R (1967) Two new methods of preventing immunity reaction in organ transplantation. German Science Re-Emerges:112

Schmidt-Mende M, Brendel W (1967) Experimentelle Untersuchungen zum Energiestoffwechsel der kühlkonservierten Niere. Z Ges Exp Med 143: 250

Seifert J, Brendel W, Pichlmayr R (1968) Normale Verteilung ^{51}Cr-markierter Lymphocyten in der Ratte und ihre Änderung nach Antilymphocytenserum. Z Ges Exp Med 146:275

Pichlmayr R, Mikaeloff P, Messmer K, Rassat JP, Bomel J, Tidow G, Fateh-Moghadam A, Junghans M, Brendel W (1968) Homologe Lebertransplantation beim Hund unter immunsuppressiver Behandlung mit Antilymphocytenserum. Z Ges Exp Med 146:61

Baethmann A, Reulen HJ, Brendel W (1968) Die Wirkung des Antimetaboliten 6-Aminonicotinamid (6-ANA) auf Wasserund Elektrolytgehalt des Rattenhirns und ihre Hemmung durch Nicotinsäure. Z Ges Exp Med 146:226

Brendel W, Land W (1968) Überraschende Ergebnisse durch intravenöse Therapie mit Antilymphozytenserum bei Organtransplantationen. Dtsch Med Wochenschr 93:2309

Brendel W (1968) Experimentelle Chirurgie in Deutschland. Med Klin 63:561

Brendel W, Müller C, Messmer K, Reulen HJ (1968) Der klinische Tod in Hypothermie. Z Ges Exp Med 146:189

Land W, Rudolph R, Brendel W (1968) In-vitro-Wirkung eines heterologen Antiserums auf Lymphozyten der Ratte: eine elektronenmikroskopische Studies. Blut 18:1

Messmer K, Devens K, Brendel W, Kraemer M (1968) Die Wirkung von Katecholamin-, Hypertensin- und Vasopressininjektionen auf die Leberdurchblutung des Hundes. Z Ges Exp Med 145:120

Pichlmayr R, Brendel W, Beck G, Schmittdiel E, Pichlmayr I, Thierfelder S, Fateh-Moghadam A, Land W (1968) Gewinnung von heterologen Immunseren gegen menschliche Lymphozyten. Klin Wochenschr 46: 249

Pichlmayr R, Brendel W, Zenker R (1968) Erfahrungen mit heterologen Antilymphozytenseren beim Menschen. Münch Med Wochenschr 110: 893

Rinecker H, Mendler N, Baethmann A, Brendel W (1968) In vivo frequency and force of dog gastrointestinal contractile activity under pathophysiological conditions. Proc Intern Union Physiol Sci 7

Trepel F, Pichlmayr R, Kimura J, Brendel W, Begemann H (1968) Therapieversuche mit Antilymphocytenserum bei Autoaggressionskrankheiten. Klin Wochenschr 46:856

Brendel W, Land W, Hopf U, Seifert J (1969) Induction of tolerance in man to horse IgG. Lancet ii:1141

Hopf U, Land W, Seifert J, Fateh-Moghadam A, Brendel W (1969) Induction of tolerance of horse *-globulin fractions in adult dogs. Br J Surg 56:617

Land W, Koczorek KR, Baethmann A, Stass P, Brendel W (1969) Immunsuppression durch Aldosteron. Pflügers Arch Eur J Physiol 312:2

Land W, Seifert J, Fateh-Moghadam A, Hopf U, Brendel W (1969) Immunological tolerance induced in adult dogs by small amounts of horse IgG. Transplantation 8:748

Land W, Rudolph R, Brendel W (1969) Ultrastrukturelle Zell-und Organveränderungen nach Verabreichung eines heterologen Antilymphozytenserums bei der Ratte. Blut 19:470

Land W, Frick E, Roscher R, Brendel W, Baethmann A (1969) Wirkung eines heterologen Antilymphozytenserums auf die experimentelle allergische Encephalomyelitis. Klin Wochenschr 47:633

Mendler N, Weishaar E, Brendel W (1969) Eine Herz-Lungen-Maschine für Ratten als experimentelles Modell der extrakorporalen Zirkulation. Thoraxchir Vask Chir 17:534

Messmer K, Brendel W, Holper K, Sunder-Plassmann L (1969) Extreme Blutverdünnung durch Volumensubstitution. Anaesth Wiederb 30:55

Pichlmayr R, Land W, Wagner E, Fateh-Moghadam A, Eulitz M, Brendel W (1969) Experimentelle Untersuchungen zur Verbesserung der Spezifität von Antilymphozytenseren und ihrer Verträglichkeit. Klin Wochenschr 47:628

Reulen HJ, Medzihradsky F, Enzenbach R, Marguth F, Brendel W (1969) Electrolytes, fluids and energy metabolism in human cerebral edema. Arch Neurol 21:517

Seifert J, Brendel W (1969) The effect of antilymphocyte serum on homologous ^{51}Cr-labeled lymphocytes in rats. Transplantation 8:806

Seifert J, Proels H, Messmer K, Land W, Hopf U, Mehnert H, Brendel W (1969) Die Glukoseresorption über den Ductus thoracicus des Hundes. Pflügers Arch Eur J Physiol 312:75

Guthy E, Brendel W (1969) Stapling devices and their use in surgery. Prog Surg 7:56

Brendel W, Hopf U (1969) Autoimmunerkrankungen. Klinik und Therapie. Schattauer, Stuttgart

Baethmann A, Brendel W, Koczorek KR, Enzenbach R (1970) Pathophysiologie des traumatischen Hirnödems. Dtsch Med Wochenschr 95:1020

Brendel W (1970) Grundlagen des Antilymphozytenglobulins und seine therapeutischen Möglichkeiten. Arzneimittel Forsch 20:191

Reulen HJ, Steude U, Brendel W, Hilber C, Prusiner S (1970) Energetische Störungen des Kationentransports als Ursache des intrazellulären Hirnödems. Acta Neurochir 22:129

Brendel W (1971) The clinical use of ALG. Transplant Proc III:280

Brendel W (1971) Gegenwärtiger Stand der Organtransplantation. Dtsch Ärztebl 68:3078

Brendel W (1971) Heterologes Antilymphocytenglobulin. Dtsch Med Wochenschr 96:1851

Hagl S, Beck G, Brendel W, Land W, Mayr N, Mendler N, Pielsticker K, Sebening F (1971) Xenogeneic heart trans-plantation in a closely related species system, fox-dog. German Med Month 1:3

Hopf U, Land W, Furthmayr H, Timpl R, Brendel W (1971) Versuche zur Induktion einer immunologischen Toleranz gegen L-Asparaginase von *E. coli*. Klin Wochenschr 49:711

Land W, Schilling A, Aldenhoff J, Lamerz R, Pielsticker K, Mendler N, Brendel W (1971) In vitro studies on the mechanism of hyperacute xenograft rejection. Transplant Proc III:888

Messmer K, Brendel W (1971) Pathophysiologische Aspekte des hypovolämischen, kardiogenen und bakteriotoxischen Schocks. Med Welt 22:1159

Pfister H, Lani K, Demmler K, Thierfelder S, Fateh-Moghadam A, Land W, Brendel W, Stich W (1971) Über die Wirkung von heterologem Antilymphozytenglobulin bei chronischer lymphatischer Leukämie. Dtsch Med Wochenschr 96:1468

Chaussy C, Hammer C, von Scheel J, Eisenberger F, Land W, Pielsticker K, Klein U, Brendel W (1972) Verhalten xenogener Nierentransplantate im nah verwandten Speziessystem. Res Exp Med 157:257

Hammer C, Land W, Pielsticker K, Brendel W (1972) Experimentelle Xenotransplantation in entfernt stammesverwandten Speziessystemen. Res Exp Med 159:124

Brendel W (1973) Antilymphozytenglobulin heute. Münch Med Wochenschr 115:241

Chaussy C, Hammer C, Eisenberger F, Pielsticker K, Brendel W (1973) Experimentelle xenogene Nierentransplantation im nah verwandten Speziessystem Katze-Hund. Urol Internat 28:230

Hammer C, Chaussy C, Brendel W (1973) Preformed natural antibodies in animals and man. Eur Surg Res 5:162

Hammer C, Chaussy C, Krebs G, Fiedler H, Pongratz H, Brendel W (1973) A new concept for classification of xenograft rejection. Eur Surg Res 160:19

Hammer C, Land W, Brendel W (1973) Experimentelle Xenotransplantation in entfernt stammesverwandten Speziessystemen. Res Exp Med 160: 239

Ring J, Seifert J, Lob G, Land W, Coulin K, Brendel W (1973) Zum Risiko einer ALG-Therapie. Mögliche Nebenwirkungen, Prophylaxe und Behandlung. Klin Wochenschr 51:487

von Specht BU, Seinfeld H, Brendel W (1973) Polyvinylpyrrolidone as a soluble carrier of proteins. Hoppe Seylers Z Physiol Chem 354:1659

Brendel W (1974) Höhenakklimatisation und Erschöpfung. Ärztl Prax 26: 2837

Heberer G, Brendel W, Schildberg FW, Feifel G (1974) Aufgaben und Organisation chirurgisch-klinischer Forschung. Chirurg 45:490

Ring J, Lob G, Angstwurm H, Brass B, Backmund H, Seifert J, Coulin K, Frick E, Mertin J, Brendel W (1974) Intensive immunosuppression in the treatment of multiple sclerosis. Lancet ii:1093

Ring J, Seifert J, Lob G, Coulin K, Brendel W (1974) Humanalbuminunverträglichkeit: Klinische und immunologische Untersuchungen. Klin Wochenschr 52:595

Ring J, Seifert J, Mertin J, Brendel W (1974) Prolongation of skin allografts in rats by treatment with linoleic acid. Lancet 1:1331

Bohmert H, Brendel W, Chaussy C, Sollinger HW (1974) Hautersatz bei Verbrennungen mit Spalthautnetz- und Xenotransplantaten. Langenbecks Arch Chir 337:238

Brendel W (1975) Prinzipien und Aufgaben chirurgischer Forschung. Ärztl Praxis 27:2399

Chaussy C, Hammer C, von Scheel J, Pielsticker K, Sollinger HW, Pfeiffer KJ, Pongratz H, Brendel W (1975) Xenogeneic skin and kidney transplants in a closely related canine system, fox-dog. Transplantation 20:150

Ring J, Seifert J, Lob G, Brendel W (1975) Immunologische Untersuchungen bei Patienten mit klinischer Humanalbuminunverträglichkeit. Langenbecks Arch Chir Suppl Chir Forum: 411

Ring J, Seifert J, Messmer K, Brendel W (1975) Untersuchungen zur Frage der Nebenwirkungen bei Anwendung von Plasmaersatzmitteln. Klin Anaesth Intensivther 9:58

Schilling A, Land W, Pielsticker K, Aldenhoff J, Brendel W (1975) Experimentelle Xenotransplantation in entfernt stammesverwandten Speziessystemen: Interaktion humoraler Faktoren bei der hyperakuten xenogenen Abstossungsreaktion (HXAR) am Ratte-Hund-Modell. Res Exp Med 165:79

Brendel W, Ring J, Seifert J (1975) Immunosuppressive treatment of multiple sclerosis with ALG and/or thoracic duct drainage. Neurology 25:490

Duswald KH, von Scheel J, Hammer C, Brendel W (1976) Langzeitüberleben von Hauttransplantaten im xenogenen System Wolf-Hund. Res Exp Med 167:255

Ring J, Seifert J, Patzelt U, Lob G, Probst J, Brendel W (1976) Beurteilung von Schweregrad und Verlauf der chronischen posttraumatischen Osteomyelitis. Arch Orthop Unfallchir 85:241

Schilling A, Land W, Pratschke E, Pielsticker K, Brendel W (1976) Dominant role of complement in the hyperacute xenograft rejection reaction. Surg Gynecol Obstet 142:29

Schmiedek P, Gratzl O, Spetzler R, Steinhoff H, Enzenbach R, Brendel W, Marguth F (1976) Selection of patients for extra-intra-cranial arterial bypass surgery by rCBF measurements. J Neurosurg 44:303

Eisenberger F, Schmiedt E, Chaussy C, Wanner K, Forssmann B, Hepp W, Pielsticker K, Brendel W (1977) Berührungsfreie Harnsteinzertrümmerung. Dtsch Ärztebl 17:1145

Ring J, Seifert J, Brendel W (1977) High incidence of horse serum protein allergy in various autoimmune disorders. J Allergy Clin Immunol 59, 3:189

Seifert J, Lenz J, Welter H, Brendel W (1978) Microcirculation and immune response in the gastro-intestinal tract of dogs. Eur Surg Res 10 [Suppl 1]:60

Brendel W, Chaussy C, Schmiedt E, Eisenberger F (1979) Berührungsfreie Zertrümmerung von Nierensteinen – eine neue Therapie? In: Alexander-von-Humboldt-Stiftung, Vorträge einer Tagung ehemaliger Humboldt-Stipendiaten in Japan, p 197

Brendel W (1979) Wird Höhenmedizin wieder aktuell? Münch Med Wochenschr 121:904

Bartmann P, Brendel W (1980) Isolierung und Nachweis antigenspezifischer T-Zell-Rezeptoren. J Clin Chem Clin Biochem 18:1

Chaussy C, Brendel W, Schmiedt E (1980) Extracorporeally induced destruction of kidney stones by shock waves. Lancet ii:1265

Wembacher J, Senftleber I, Hammer C, Brendel W (1980) Lymphocytes with suppressive activity isolated from dog renal allografts under rejection. Immunobiol 157:296

Land W, Hammer C, Weiss EP, Hofmann G, Brendel W (1980) Spleen transplantation in the dog as a model for studying immunological feedback reactions. Immunobiol 157:240

Enders G, Seifert J, Hauffe A, Brendel W (1980) Immune response to an orally-applied antigen after removal of the Peyer's patches. Immunobiol 157:216

Brendel W (1981) Nierensteinzertrümmerung mit Stosswellen. Umschau 81:100

Welter HF, Krause H, Hammer C, Brendel W (1981) Xenogene und allogene Bluttransfusion vor xenogener Nierentransplantation im System Fuchs-Hund. Langenbecks Arch Chir Forum 81:199

Endrich B, Laprell-Moschner C, Brendel W, Messmer K (1982) Effects of prolonged cold injury on the subcutaneous microcirculation of the hamster. Res Exp Med 181:49

Brendel W (1983) Die Bedeutung der experimentellen Chirurgie. Fortschr Med 101:101

Brendel W, Enders G (1983) Shock waves for gallstones: animal studies. Lancet i:1054

Hammer C, Land W, Stadler J, Koller C, Brendel W (1983) Lymphocyte subclasses in rejecting kidney grafts detected by monoclonal antibodies. Transplant Proc XVI:356

Lersch C, Hammer C, Ganghoff O, Brendel W, Nath G (1983) Infrared contact coagulation. A new therapy for solid tumors. Naturwissenschaften 70:416

Chaussy C, Schmiedt E, Jocham D, Walter V, Brendel W (1983) Stosswellentherapie zur Behandlung von Nierensteinen. Münch Med Wochenschr 125:151

Brendel W (1984) Chirurgische Forschung. Münch Med Wochenschr 126: 1263

Brendel W (1984) Nierenlithotripsie. Inn Med 11:197

Brendel W (1984) Experimentelle Chirurgie. Münch Med Wochenschr 126: 819

Brendel W (1984) Stosswellen. Ein neues therapeutisches Prinzip in der Medizin. Münch Med Wochenschr 126:3

Hofmann GO, Land W, Weiss EP, Hammer C, Permanetter W, Brendel W (1984) Spleen transplantation in the dog as a model for studying immunological feedback reactions. Eur Surg Res 16:40

Brendel W, Delius M, Enders G (1985) Experimental destruction of gallstones by shock waves. In: Paumgartner G, Stiehl A, Gerock W (eds) Enterohepatic circulation of bile acids and sterol metabolism. Falk Symposium 42. MTP Press, Lancaster, p 383

Hobbhahn J, Vogel H, Knothe N, Brendel W, Peter K, Jesch F (1985) Hemodynamics and oxygen transport after partial and total blood exchange with pyridoxalated polyhemoglobin in dogs. Acta Anaesthesiol Scand 29:537

Krombach F, Schödel F, Lersch C, Hammer C, Brendel W (1985) Granulocytenaktivierung bei Nierentransplantatabstossung. In: Stelzner F (ed) Chirurgisches Forum '85 für experimentelle und klinische Forschung. Springer-Verlag, Berlin Heidelberg; New York, p 95

Lersch C, Thiel M, Hammer C, Schuke M, Krombach F, Brendel W (1985) Splenic fine needle aspiration cytology in small animals with transplanted tumors. Transplant Proc XVII:2132

Lersch C, Schreiner J, Demmel N, Hammer C, von Finckenstein J, Koller J, Dirschedel P, Krombach F, Günther B, Brendel W (1985) Monitoring immunocompetent cells in the peripheral blood of stomach cancer patients after splenectomy and gastrectomy. J Cancer Res Clin Oncol 110:225

Schödel F, Krombach F, Lersch C, Hammer C, Brendel W (1985) Chemiluminescence of polymorphonuclear leukocytes during renal allograft rejection. Transplant Proc XVII:2554

Brendel W (1986) Shock waves: a new physical principle in medicine. Eur Surg Res 18:177

Brendel W (1986) Experimentelle Voraussetzungen der Organtransplantation. Münch Med Wochenschr 128:875

Dibelius A, Königsberger H, Walter P, Permanetter W, Brendel W, von Specht BU (1986) Prolonged reversal of diabetes in the rat by transplantation of allogeneic islets from a single donor and cyclosporine treatment. Transplantation 41:426

Dibelius A, Königsberger H, Permanetter W, Walter P, Brendel W, von Specht BU (1986) Prolonged pancreatic islet allograft survival by cyclosporine compared to whole pancreas allograft survival. Transplant Proc XVIII:1167

Hart H, Seifert J, Brendel W (1986) Wirkung und Nebenwirkung von Taurolin bei einer experimentellen Peritonitis der Ratte. Langenbecks Arch Chir 368:149

Knorr-Held S, Brendel W, Kiefer H, Paal G, von Specht BU (1986) Sensitization against brain gangliosides after therapeutic swine brain implantation in a multiple sclerosis patient. J Neurol 233:54

Krombach F, Schödel F, Rath M, Gokel M, Lissner J, Brendel W (1986) Renal transplant rejection in core biopsy, fine needle aspiration cytology and magnetic resonance imaging: a comparative study. Transplant Proc XVII:1099

Lersch C, Hammer C, Thiel M, Lenz P, Brendel W (1986) Analysis of tumor-enhancing lymphocytes in murine spleens removed at different time intervals after 3 LL-tumor inoculation. J Exp Clin Cancer Res 5:179

Reichenspurner H, Ertel W, Reichart B, Kempkes BM, Jamiesen SW, Hammer C, Brendel W (1986) A new model for heteroorthotopic heart-lung transplantation in the dog. J Surg Res 41:301

Sauerbruch T, Delius M, Paumgartner G, Holl J, Wess O, Weber W, Hepp W, Brendel W (1986) Fragmentation of gallstones by extracorporeal shock waves. New Engl J Med 314:818

Schödel F, Krombach F, Hammer C, Brendel W (1986) Peripheral blood and intrarenal phagocytic chemiluminescence during acute kidney graft rejection. Inflammation 10:271

Teichmann R, Pratschke E, Grab J, Hammer C, Brendel W (1986) Gastrin release by interleukin-2 and γ-interferon in vitro. Can J Physiol Pharmacol 86:62

Brendel W, Delius M, Goetz A (1987) Effect of shock waves on the microvasculature. Progr appl Microcirc 12:41

Brendel W, Delius M, Enders G (1987) Gallenblasensteinzertrümmerung durch extrakorporal induzierte Stoßwellen. Z Gastroenterol [Suppl 2] XXV:22

Delius M, Enders G, Heine G, Stark J, Remberger K, Brendel W (1987) Biological effects of shock waves: lung hemorrhage by shock waves in dogs – pressure dependence. Ultrasound Med Biol 13:61

Goetz A, Deininger FD, Conzen P, Brendel W (1987) Microcirculatory effects of LTC_4 and LTD_4 in the skin fold preparation of awake hamsters. Prog Appl Microcirc 12:307

von Specht BU, Strigl G, Ehret W, Brendel W (1987) Protective effect of an outer membrane vaccine against *Pseudomonas aeruginosa* infection. Infection 15:408

Königsberger H, Dibelius A, Permanetter W, Walter P, Brendel W, von Specht BU (1987) Influence of postdiabetic onset time and immunosuppressive treatment of islet grafts in the spontaneous diabetic BB/W rat. Transplantation 44:358

Brendel W (1988) Aspekte und Ergebnisse experimenteller Chirurgie. Münch Med Wochenschr 130:203

Enders G, Ruckdeschel R, Teichmann R, Brendel W (1988) Changes of immunoglobulin concentrations in the bile after proximal gastric vagotomy in rats. Scand J Gastroenterol 23:301

Delius M, Enders G, Zengrong X, Liebich HG (1988) Biological effects of shock waves: kidney damage by shock waves in dogs – dose dependence. Ultrasound Med Biol 14:117

Delius M, Jordan M, Eizenhoefer H, Marlinghaus M, Heine G, Liebich HG, Brendel W (1988) Biological effects of shock waves: kidney haemorrhage by shock waves in dogs – administration rate dependence. Ultrasound Med Biol 14:689

Conzen P, Hobbhahn J, Goetz A, Habazettl H, Granetzny T, Peter K, Brendel W (1988) Splanchnic oxygen consumption and hepatic surface oxygen tensions during isoflurane anesthesia. Anesthesiology 69:643

Brendel W, Baethmann A (1989) Chirurgische Forschung in München. Rückblick und Ausblick. Münch Med Wochenschr 131:505

Wilmer A, Gambihler S, Delius M, Brendel W (1989) In vitro cytotoxic activity of lithotripter shock waves combined with adriamycin or with cisplatin. J Cancer Res Clin Oncol 115:229

A Very Decided Preference

Jean Medawar

This 20th jubilee makes me realise how much you have done, Walter, not only for Peter and me, but for everyone else, especially your students. You picked them, I understand, not only because they were bright, but because they could sing and because they could ski. I had my first chance of falling about in the snow here, and it was wonderful. And now it is lovely to be asked back again.

Some people will wonder why I have been asked, because I am neither a surgeon nor an immunologist. But, as you may know, I was married to Peter Medawar for 50 years, so I certainly know something about him; that is the reason why I have been invited. Walter wanted you to know a little bit about what he was like, and what sort of attitude to life he had, in case it might be useful to you too. So when Mrs. Stein telephoned me and asked what I would speak about, I said "A very decided preference". I had better explain that phrase. It is the title of a book that an American publisher asked me to write about 50 years with Peter.

What was he like? He was not just a great man; I think he was three great men. He was a great scientist, he was a hero, and he was also a great writer. So this title "A very decided preference" describes what my preference was; it also describes a phrase that he used in the last third of his life. For two-thirds of his 72 years he was very strong and upright and energetic, and one felt he could do almost anything, and he felt it too. Then he had a stroke which was big enough to kill him, but it did not. People used to asked him "How do you manage to be so cheerful, with only half your eyesight and no left hand or arm and the leg in a caliper or splint?" It was so heavy that when I took it off at night, I wondered how he could walk at all. He would answer: "I have a very decided preference for remaining alive." And of course that is – as any doctor or surgeon knows – the best thing for getting you on your feet again; if you want to stay alive, you are highly motivated to do so, and you do everything you can to stay alive.

I have remembered many stories about Peter for the book that I have written for the publisher W.W. Norton; every story illustrates his fantastic ebullience and love of life and learning, the extraordinary breadth of his reading and the use he made of it. In these few minutes all I can do is give you a sort of flavour of what he was like, a very brief sketch of his life and the attitude that he had towards life – both when he was in the pink of health and energy, and then when he was dragging along after the stroke and unable to

K. Messmer M. Stein (Eds.)
Pathways in Applied Immunology

see anything on his left side. When he was addressing an audience a third of it on the left just was not there, and when he read, the margin on the left was not there either. I hope that I have said enough so that people who did not know him may have an idea of what he was like, and why Walter has spoken so warmly of him.

He was born in 1915. He was a perfect example of hybrid vigour. His mother was a tall, very plain, very wonderful English woman whom he adored. She had a terrific sense of humour, and she adored him. His father was a short, not highly educated Lebanese businessman; so the mixture was very hybrid. He was born, by chance, in Rio de Janeiro, and this was a great handicap, not just to his mother who told me that the hospital she was in (and this gives you an idea how things have changed in 50 years) in Petropolis had no sheets; they had blankets, and the blankets had fleas, and they bit her unmercifully. The only nursing care was relatives – you had to bring in your relatives, and if you had no relatives, it was tough. Peter grew up speaking a little Portuguese. In Rio he began his lifelong love of opera when his father gave him a season's ticket to the opera house. He also adored gambling. In his book *Memoirs of a Thinking Radish* he describes how his father used to reporach him for gambling, saying it would do his reputation no good at the Chamber of Commerce, and Peter replied: "Father, if I came away from it with empty pockets, that might do you some harm at the Chamber of Commerce; but as it is, I emerge with my pockets full and with a satisfied expression!" He nearly always won. His mother very sensibly thought that he should have an English education. So the poor little boy had to go to a ghastly school in Broadstairs on the South Coast of England. It was run by a former army captain who soon absconded with all the school funds. So Peter went on to another school. There, he told me, the headmaster was a persecution maniac; he used to sit on Peters' bed and tell him that the housekeeper was after him with a carving knive. I think it shows how sanguine his nature was that these things appeared to have no deleterious effect on his character (of course analysts might say that he was hiding the damage).

He was very sanguine and very intelligent. At this time he began to read ferociously. He read everything he could lay hands on: cycling magazines, *Beano* and *Rainbow* (English children's comics) and the *Encyclopedia Britannica*. In due course he went to a public school (which means an expensive one) called Marlborough. And there were cold baths and beatings – it sounds awful. But Peter said, and this is another example of his sanguine nature, that the advantage of this for him was that afterwards nothing could ever be so bad again. I found that a very helpful idea. He had great good luck at Marlborough, because the science master, Ashley Gordon Lowndes, was a real scientist. He taught science at Marlborough, and he was delighted when he got hold of an intelligent child; he realized that Peter had enormous potential. If Peter or another boy irritated him, he would knock their heads together – literally! The boys knew it was only because he had been wounded in the war and had a metal plate in his scalp. So they laughed off his bad temper. But he taught Peter hard and so well that he got into Oxford University when he was

17. He went to Magdalen College, a wonderful college, and he met me in his second year. He said that this was very fortunate for him, because I was moved by the whole idea of scientific inquiry. I had just read Rene Valery Radot's *Life of Pasteur* and felt I wanted to dedicate myself, and that was just what he wanted. He once said that if he had not met someone who understood him, and if he had not found science, he thought he would have had a nervous breakdown, because he had so much energy and such ferocity of purpose, that it had to find the right vehicle. We soon realized that we wished to spend our lives with each other. I must tell you, because this is a sign of how things have changed in my lifetime, that when I wanted to borrow one of Peter's essays, a fellow student said to me: "I suppose you knew he's not quite English?" "He's half Arab!" To him this was something from which he should protect me. I did not pay the slightest attention. Before we married Peter said, half in fun: "You don't want to tell me your dreams at breakfast, do you?" I said no, and he said: "You don't want to write a novel, do you?" And I again said no, and he replied: "Then that's alright."

I do not know how a girl like me, brought up with very well groomed, carefully selected young men in Cambridge, ever understood that under his careless appearance he was full of fire and light. But I did. At the beginning of his last term at Oxford he said to me: "Would you like me to get a First?" I said "Well, of course." He began to revise; he selected what he knew he would not be asked because it had been asked the year before, he selected what he knew he would have to know as a zoologist, and he selected what he knew he was superbly good at. He got his First class degree and he got a junior fellowship at Magdalen College and began research. For the first two years he did not focus – he tried to, but the problems that he was given were not big enough. Then one day there was a terrible accident: We were in the garden one Sunday in 1940, and we had been warned that we might be attacked by low flying planes. I can still see the plane coming towards us over the trees at the end of the garden. I thought "This is it!" and I scooped up our two-year-old daughter and ran into the air-raid shelter; shortly after that there was a crash that shook the ground; it was an English plane that crashed, and one of the airmen in it was appallingly burned – two-thirds of his body had third-degree burns. Peter was asked to come and to see if he could help. At that time he had been trying to work out whether there were any factors which would promote connective tissue growth. When he saw this airman, his attention became fixed on the solution of this crucial problem. It was a question whether the man would live or die, and what he could do about it. Is it possible to culture skin? Is it possible to make up a soup of skin cells and spread them on? What could be done? Thereafter Peter concentrated on why it is not possible to transplant skin from one person to another.

At this time he said: "Once you have tackled a problem and made any headway in solving it, you are hooked, and no other kind of life will do." He also said: "How lucky I am to be paid for what I would like to do even if I wasn't paid." Of course, there were a lot of things that he had to do that were not so amusing, but from then on his life was in science. Later on he added the administrative burden of being a professor (at age 33) in Birmingham, and after 3¾ years, in London. While he was in Birmingham, he and Leslie Brent and Bill Billingham spent a lot of time in a cattle shed at Cold Norton Farm trying to graft skin from one cow to another. I need not go into that because you know it, but it was the beginning of finding that the problem he was attacking, why skin cannot be grafted between people, belongs to the field about which something was known – the field of immunology. This discovery gave a tremendous boost to surgeons who were longing to be able to graft skin from a donor to a burnt or injured patient.

When he settled in London in 1953, the era of what we called the supermice began. He had the idea of giving an injection of skin through the body wall of a pregnant mother into each of the embryos; when they were born he hoped they would accept this dose as part of themselves instead of reacting to it as something foreign. The mice were born, and when they had grown their fur coats, Peter grafted a small patch of skin from the mouse that had given the injection of cells – and the grafts stayed on. These mice were known as the supermice. They were very smart; they lived in a big glass bowl with sawdust at the bottom, and they learned that if they stood and pushed the lid up, all together, they could get it off. One of them jumped at the right moment, scuttled across the floor and popped down the hole in the floor, made by the gas pipe and into the space under the floor boards, where a wild mouse had his territory. The noise of the fighting could be heard above. They did everything they could: they put a little female mouse in a cage by the hole, and they offered bread and cheese and nuts, but nothing worked until the battle was ended. Then the supermouse dragged himself out. He had lost about one-third of his body weight. They treated him like a head of state, gave him brandy and milk, but nothing worked, and he died a sort of hero's death. The other supermice survived, and the principle of immunological tolerance was born.

In London Peter served on many government committees, gave lectures and wrote ten books; for leisure he played cricket, chess and bridge and listened to music. He had a tremendous sense of order. He would not even go to bed without putting out the next day's clothes for the morning, so that he should not waste any time choosing them. He hated shaving; he said "God, I'm so bored with this every morning!" Then I got him an electric razor, and he thought that was a bit better, but even then he resented the time away from thinking or doing. His sense of order was attached above all to the hope of progress. He was a tremendous meliorist (that comes from Latin meaning 'better'); a meliorist is midway between an optimist, who expects everything to be all for the best, and a pessimist who expects everything to be consistently bad. A meliorist believes that if we do the right thing, we may be able to make things better. Thank God he had those qualities, because he needed them. By 1969 he had been a Fellow of the Royal Society for 20 years, and I have been told that he was the most likely candidate for the next president. I remember that he said: "I don't mind if someone else is elected, but it would be rather wonderful to follow Isaac Newton!" He had been awarded the Nobel Prize 9 years earlier for his work on acquired immunological tolerance. That year he was the President of the British Association. The President gives a public lecture at the end of the proceedings. I think that the lecture he prepared is the most wonderful, magnificent piece of writing that I have ever read or hope to read. In it he encapsulated, in marvellous language, sometimes funny, his whole outlook on life, and when he finished, one could not hear a sound for several seconds before the applause began. But, 24 hours after, in the cathedral at Exeter, whilst he was reading the lesson he had chosen from the Wisdom of Solomon, his voice began clouding, and he became unconscious. The stroke was enough to kill most people, but his preference for remaining alive helped him to recover.

The 11 years after that were spent gradually improving, working and enjoying life again. When people asked Peter "How did you feel when you were so critically ill?" he said "I myself, habitually sanguine, considered and dismissed the prospect of dying." In the 11 years after that awful stroke I never knew him to bewail his fate. He said "I neither curse God for depriving me of the use of my left arm and leg nor thank him for having spared me the use of my right." We managed well until 1980; in March of that year he had another stroke, in the brainstem, and he had to start all over again. We used to do voice exercises, roaring like lions or hissing like snakes. First his voice was a whisper, but gradually it got strong enough for him to lecture again, and that meant a lot to him. We travelled a great deal; he was invited everywhere. We had another 5 years, and then, in 1985, he had a terrible stroke; the neurologist told a friend of mine that he thought that this was Peter's last illness, but it wasn't. Two months after that Peter said to me: "You know, when you've been *there* and back" (meaning to the river Styx), where death with a scythe is waiting to receive you, people look at you with a new respect." I was so pleased at him being able to have this concept. I thought I would be rather earthy and asked casually "Did you meet that old bugger with the scythe?" And he said "Yes!" So I continued "What did you say to him?" and Peter answered "He said: *avaunt*!" – that means in Shakespearian English "Get the hell out of here!" Then I asked "Well, go on, what was he doing?" and Peter said: "He was picking his teeth with the scythe." He spent a year in hospital, and for 14 months, after he came home I looked after him, and we managed until October 1987. We celebrated our golden wedding anniversary in February 1987.

When he died, I asked that instead of sending flowers people should give a donation to the National Trust for trees, so that we could have a Medawar Grove at Alfriston in Sussex, where our grave is, in a most beautiful place by the sea. We have enough money to plant a Medawar Grove, and I plan to add a bench next year. When you come to England you may like to visit Alfriston, a beautiful village on the river. On the gravestone I have had carved the last sentence of the passage that he gave in the Presidential Address to the British Association. It was written by the philosopher Thomas Hobbes, and Peter loved it: "There can be no contentment but in proceeding".

When Peter finished the Address he added "I agree" – and so do I.

Reflections

L. Brent

Introduction

Walter Brendel suggested that contributions to this 20th symposium, which is clearly rather special, might be less formal and include a smattering of personal reminiscences and nostalgia, and correspondingly less conventional science. If the programme is anything to go by, I am probably the only one who has taken this advice seriously – you may well think *too* seriously! What I wish to present is indeed a mixture of all these. They include personal reflections on my scientific life, with an emphasis on missed opportunities rather than on whatever successes may have come my way, and on some of the more risible moments – to quote from Peter Medawar's *Memoirs of a Thinking Radish* [1] when reviewing his own life.

But first a few words about these symposia. In 1969 I received an invitation to what I think was the first meeting of this series. I was in a bit of quandary, for I had just left Southampton to take up the Chair of Immunology at St Mary's Hospital Medical School, and I was faced with the task of rebuilding a department that had been very flourishing under my illustrious predecessor Rodney Porter, but that had been allowed to become defunct in the 2 years after his move to Oxford.

Perhaps I should say a few words about Avrion Mitchison here, for it was Av who was a very important influence in this move. I had been in Southampton for only 4 years, and I was guilt-ridden at the thought of moving so soon from my post there when I was invited to St Mary's. It was at a meeting over dinner that Av encouraged me to take the plunge, and I have been grateful to him for that ever since.

The invitation from Walter Brendel and Konrad Messmer, with its mix of immunology and skiing, struck me at that time as a trifle frivolous, and, foolishly, I declined it. After all these years I think I can add that there was another reason that made me turn down Walter's invitation. Since my enforced exodus from Berlin late in 1938, without my family, I had not been back to Germany except as a soldier in the immediate post-war period and when attending a medical congress in Berlin in 1965; I clearly still had many complex feelings to work through so far as Germany was concerned. Walter and Konrad, fortunately quite undeterred, sent me an invitation to the 2nd symposium, and having been primed by Gene Lance, I accepted. With the

K. Messmer M. Stein (Eds.)
Pathways in Applied Immunology

exception of the year in which I was President of the Transplantation Society, when I was heavily preoccupied with preparations for the Rome Congress, I think I have attended all the others. I am enormously grateful to Walter and Konrad and to Claus Hammer for these invitations; each has been greatly cherished and valued, and the meetings have given me enormous pleasure – not just for the science or the skiing, although the two have acted in a mysteriously synergistic manner, but quite as much for the opportunity of forging friendships with young German scientists and with some who were not quite so youthful at the time, for example, Walter and Jutta Brendel and Mechthild Stein. So, quite apart from the joy of discussing immunology in such congenial surroundings and the fun and games on the ski slopes, these meetings have provided me with some much needed therapy. They have helped me to look at the modern Germany with a degree of objectivity that might otherwise not have been possible. In my talk at Walter's moving 65th birthday celebrations last year – and I felt very privileged to have been asked to take part – I said that "It has always struck me as notable that among the hard core of immunologists and transplant surgeons invited to these meetings from abroad there were rather more Jews that might be expected from random chance. Now, I feel that this tells us something very important about Walter Brendel." And so I believe it does. I am greatly in Walter's debt for having given me the opportunity of attending these marvellous meetings.

First Experiments and Early Years

I became Peter Medawar's research student in 1951, or strictly speaking it was January 1952, because Rupert Billingham and I were left in Birmingham for a few months whilst Peter prepared the laboratories and animal house at University College, London, where he had been appointed to the Chair of Zoology. It was Rupert Billingham ("Bill") who had suggested to Peter that I might make a suitable research student, at a time when I was heading in a totally different direction, for I had decided to become a school teacher. I have no cause for complaint at this change of direction; indeed, I am eternally grateful to Bill for having taken the initiative over this and to Peter for taking me on, for my years with Peter – all 14 of them – were productive and exciting and memorable in every way. There can be no doubt at all that he was an extraordinary and unique man, as we have heard from Jean.

To help fill the 3 months' gap, Bill drew my attention to a brief paper published by a South African pathologist. In it he described a higher growth rate of allogeneic tumors in outbred mice that had been treated with the vital dye trypan blue, and I decided to see whether this applied equally to skin allografts in rabbits – my very first experiments in Birmingham. Substantially prolonged survival was observed, and although I was very concerned at pumping my rabbits full of trypan blue – and they were jolly blue, I assure you – the regeneration and epithelial growth that took place did not suggest that I was dealing with a stress phenomenon. Upon moving to London I became

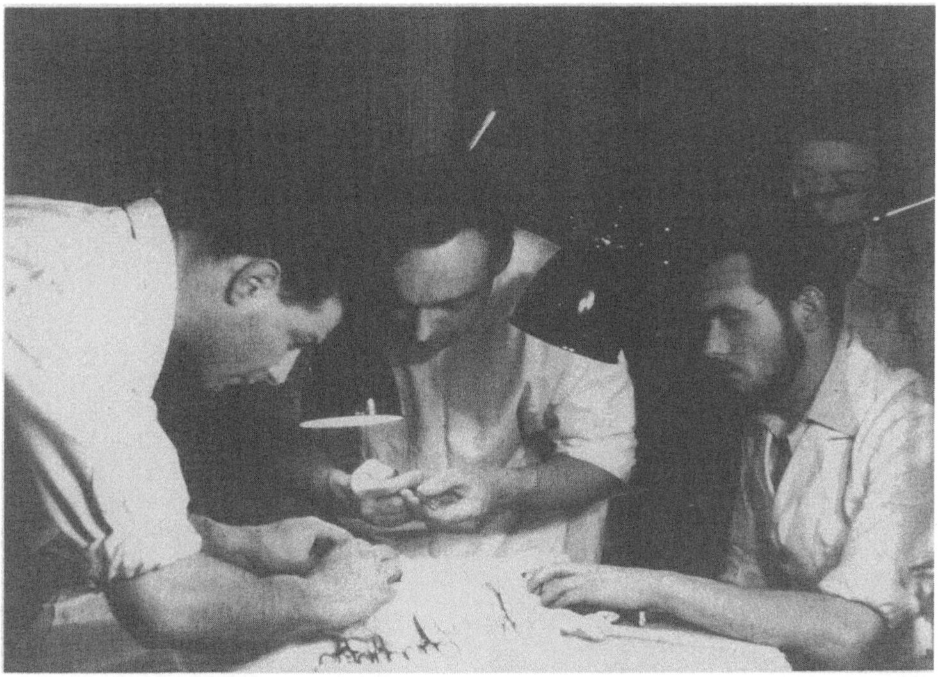

Fig. 1. Intravenous inoculation of newborn mice at University College, London, 1956: L. Brent (*left*) R.E. Billingham (*centre*) and T. Courtney (*right*)

immersed in the tolerance studies, and my data were never written up. Had I continued with that approach, and had I ascertained that macrophages were directly involved in the induction of this non-specific unresponsiveness, I might have anticipated events, i.e. the role of macrophages in immune responses, by at least a couple of decades. Well, I did not.

Bill stayed on in London for 6 years before going to the United States; in those days we were known by some Americans as "the Holy Trinity". It was clear enough who represented the Father and Son, but I never thought that Bill was much of a Holy Ghost! When I wrote my Ph.D. thesis in 1954, it was on tolerance and related questions, and I was able to include virtually every important reference in the field of transplantation immunology. Quite a few came from Medawar's own stable, which made life a lot easier. With a brilliant intellect such as Peter's it is quite easy to underestimate the contribution that Bill made to their and our collaborative investigations. Peter certainly did not, and he recognized that Bill was an exceptionally able and ingenious scientist. As Bill was the only photographer among us, I do not actually have any slides of the three of us together, but Fig. 1 shows Bill and myself working on the intravenous injection of neonatal mice with viable cells, a technique [2] that enabled us to analyse tolerance very much more easily and to recognise runt disease (or graft-versus-host disease). The third person shown was our tech-

Fig. 2. P.B. Medawar (*right*) and L. Brent on the way to visit J.F. Loutit's Radiobiology Unit at Harwell, Berkshire, 1956/1957

nician, Trevor Courtney. Figure 2 shows Peter Medawar and myself on our way to Harwell to visit John Loutit and his colleagues, who were at that time working on secondary disease after whole-body irradiation and bone marrow transplantation in mice. We put it to them that secondary disease was quite likely caused by a graft-versus-host reaction, which was at that time (1956/7) a new concept.

Caltech and Lymphocyte Receptors

I will now tell you of another failure on my part to anticipate events, this time through a lack of follow-up combined with a reluctance to publish the same data twice. In 1956 Peter arranged for me to spend a year at the California Institute of Technology with Ray Owen as a Rockefeller Research Fellow. Ray was the man who had described the red cell chimerism in cattle dizygotic twins as early as 1945, a critically important milestone in the discovery of immunological tolerance. At Caltech I met Jim Berrian who had just left the

United States Navy, and I persuaded him to join me in an enterprise designed to show whether there were "cell-bound antibodies" on lymph node and spleen cells. This study was unfortunately published in 1958 in a volume of the New York Academy of Sciences Proceedings [3]; it never reached the right people and was ignored. The protocol was as follows. Mice were sensitized with skin grafts, had their spleens and lymph nodes removed, and viable cell suspensions were prepared from them. These were exposed in vitro to aliquots of a semi-soluble extract of transplantation antigens prepared from spleens of mice from the skin donor strain. After recovery of the extracts following sedimentation of the cells, their antigenic potency was determined by assessing their ability to sensitize normal mice syngeneic with the cell donors against skin grafts of the antigen donor strain. If cell-bound antibodies, which we would now call antigen-specific receptors, were present they could be expected to bind the antigens in solution and sequester them, leaving the extract antigenically less potent than control aliquots incubated with normal cells. That is indeed what happened, and in our discussion we stated that the hypothesis most attractive to us was that the process of active sensitization had led to the generation of cells with specific membrane-bound antibodies, and that these antibodies play the dominant role in allograft rejection. On returning to Britain at the end of my sabbatical year I became involved with other projects, and I did not follow up these experiments. This work could nevertheless be regarded as a primitive and not very rigorous forerunner of much later studies on the interaction between antigens and lymphocyte receptors.

Allograft Rejection and Delayed-Type Hypersensitivity

A study that pleased me in the 1960s was the demonstration that the allograft response can be shown to have an uncanny resemblance to the tuberculin reaction, suggesting that allograft rejection is a special kind of delayed-type hypersensitivity reaction. It was a suggestion that I had made in a review in 1958 [4], and later Jean Brown and I, with Peter Medawar's support, subsequently carried out experiments in guinea pigs to test it. Provided that one uses the guinea pig, which displays cutaneous delayed hypersensitivity reactions especially well, and provided that the right kind of protocol is used (i.e. a semi-soluble antigenic extract administered intradermally), allograft responses can indeed take the form of typical delayed hypersensitivity skin reactions [5]. This notion fell into disfavour when cytotoxic T-lymphocytes (Tc) were discovered and graft rejection was mainly attributed to them. I was therefore delighted when it was convincingly revived by Ian McKenzie and his colleagues in Australia, and the consensus now is that delayed-type responses as well as T cells play a role in allograft rejection.

Suppressor T-Lymphocytes in Allograft Unresponsiveness

My group at St Mary's was one of the first to demonstrate convincingly the existence of suppressor cells in the context of allotransplantation, though to satisfy the purists I should perhaps refer to the presence of T-lymphocytes with a functional activity that causes specific suppression. In 1975 [6] we took mice that had been made specifically unresponsive in adult life with extract donor strain antigen and anti-lymphocyte serum and transferred their spleen cells to syngeneic mice treated with a limited course of anti-lymphocyte serum or sublethal X-irradiation. This procedure conferred specific unresponsiveness very effectively on naive animals, which displayed long-term unresponsiveness to donor strain skin allografts. Tolerance was not transferred when the spleen cells were T cell-depleted with an anti-Thy 1.2 antibody and complement, and we later showed that purified T cells transferred the unresponsiveness. In 1980 the Medical Research Council unfortunately declined to fund a research proposal of mine to study specific suppression further and, discouraged by my first ever rejection of a grant application, I turned to other problems. There is a moral for young scientists here: do not trust the judgement of grant-giving bodies; try somewhere else if they turn you down. This I should have done, but I didn't. T suppressor cells in experimental transplantation models and in clinical transplantation have been all the rage in the 1980s, even though there is still some doubt as to their existence as a separate subpopulation.

Anecdotes Amusing, Exciting, Painful, Embarrassing, Bizarre and Disappointing

I will now be frivolous and briefly describe, in this order, the most amusing and the most exciting, embarrassing, painful, bizarre and disappointing moments of my scientific life. One of the most amusing was the famous late night Tennerhof party in Kitzbühel organized – if that is the right word – by Richard Batchelor and myself many years ago. We had invited a few friends – we thought they were friends – to our room, but the message had spread, and something like 15 or 16 packed into our room – on (or under) beds, on chests of drawers, littering the floor, all over the place. Ruggero Ceppellini was in absolutely sparkling form in a way that I have never seen him before or since: he was wildly expansive, reminiscing about his wartime experiences as a British prisoner of war as well as about his scientific life, and he had us in paroxysms of laughter. It was wonderful to see him, for once, so carefree. Other members of the party included Konrad Messmer, Vincent Eijsvoogel, Fritz and Marilyn Bach, Mechthild Stein, Jon van Rood, Hans Balner and Ekkehard Albert, all more than slightly inflamed by our duty-free alcohol and all in excellent voice. The party became a little noisy and exceedingly amusing; what proved to be far from amusing was facing Walter Brendel's wrath the morning after. A complaint had been made by one of the guests who was not a Symposium participant (I rather think he had been involved in some sort of contretemps with a participant in a totally different context) and Walter was mightily displeased: Zeus in an ungodly rage. Room parties (at least of this kind) have not taken place to my knowledge since!

The most exciting moment was when Peter Medawar called me into his office one day in 1960 in a state of high excitement, saying "How would you like to win the Nobel Prize, Leslie?" I replied that I would indeed rather like it! Peter had just received a phone call from a journalist in Stockholm with the first intimation that he and MacFarlane Burnet had been thus honoured.

The most painful, in a physical sense, was falling off a sledge some 8 years ago in Axams. I had the (dubious, as it turned out) privilege of sharing a sledge with Mechthild Stein, rather late at night, only a trifle high on schnaps (alas, for we would have fared rather better had we drunk a little more). Our combined weight was such that we were travelling at twice the speed of sound; Mechthild has tended over the years to blame my own weight and lack of tobogganning technique and, aspiring to be a gentleman, how could I deny that? Rather than crashing into the sledge ahead of us we decided to abandon ship and threw ourselves off. The result was a torn posterior cruciate ligament. Very painful indeed!

Two events compete for the accolade of most embarrassing moment. The first was when, at the New York Transplantation Congress in 1976, Jean Hamburger introduced me as President-elect of the Transplantation Society in a very gracious and generous speech, in which he said how disgraceful he thought it was that Rupert Billingham and I had not been included in the 1960 Nobel Prize. I could not very well ignore that in my speech and said that I – and so far as I knew Bill (who was in the audience), too – did not feel in the least hard done by. I had never discussed this with Bill and I was relieved to find subsequently that he concurred. I must confess that I have, nevertheless felt that the decision to exclude *him* had been less than just. The second embarrassing moment was the loss in 1954 – through no fault of mine – of two experimental notebooks packed in a suitcase stolen from a friend's car parked close to Lichfield Cathedral on Christmas Eve, when travelling north to visit my future parents-in-law. I informed Peter and Bill right away, and they were absolutely heroic (or should I say stoic?) in concealing their concern, which must have been very acute, and they were unbelievably kind to me in the circumstances. Fortunately it turned out that most of the data (mainly on the adoptive transfer of sensitivity and on tolerance) had either already been published or were embedded in my Ph.D. thesis that I was writing at the time, so that no great harm was done. But at the time I was absolutely distraught – only partly alleviated by Peter and Bill's admirable self-control and compassion. The loss of my only two suits seemed, by comparison, a triviality, even to an impoverished Ph.D. student.

The most bizarre moment? I think three are worth recalling. The first concerns the theory of Lamarck, on which Peter had written a critical and dismissive essay many years ago. In 1979 a young Australian immunologist, Ted Steele, published a monograph in which he proposed a mechanism to account for Lamarckian inheritance of acquired characteristics, and in 1981 he and R.M. Gorczynski described some experiments on immunological tolerance that they thought proved the point [7]. They had taken tolerant mice, bred from them and shown that the offspring were specifically tolerant in that they were incapable of generating normal cytotoxic in vitro responses to donor strain antigens. Liz Simpson and I were extremely intrigued about this at the time. Liz, being a good scientist, had an open mind on the subject, but I was sceptical because Peter, Bill and I had previously done experiments in the 1950s that seemed to rule out this possibility. Nevertheless, egged on by Liz and by the thought that we might conceivably have been wrong, and with the active support of Lee Rayfield and Philip Chandler, we embarked on an extensive series of experiments in an attempt to confirm that the progeny of tolerant mice are genetically unable to produce a cytotoxic response to the donor strain antigens. The results were negative and duly published in *Nature* [8,9].

Meanwhile, whilst we were still working on this project, Steele had been offered a place by Peter and Liz in their department at the Clinical Research Centre to enable him to confirm and extend his work. In offering him a place they showed commendable open-mindedness. Producing negative data in his presence, whilst he was getting a great deal of publicity (one of the Sunday papers published a full-page article with photographs of Darwin, Lamarck and Ted Steele), was quite trying. After the publication of our negative findings the editor of *Nature*, John Maddox, arranged an open BBC radio debate between Ted and myself, with Jonathan Howard, the immunogeneticist who had just written a book on Darwin, as an impartial participant and with John Maddox in the Chair. In the middle of the discussion, which was being recorded, Steele held up a piece of paper and said "I have re-worked the data produced by Brent and his colleagues and my statistical analysis shows that they in fact support my hypothesis. . . . Their own original conclusions are totally mistaken" (or words to that effect). There was a deathly hush, the recording was interrupted, and pandemonium broke out. Although my astonishment was total, I agreed to take a look at Steele's analysis of our data, only to find that he had evidently overlooked the fact that male and female mice show slight but statistically significant differences in the production of cytotoxic T cells and, by lumping the two sexes together, he had produced misleading statistical data. The discussion was continued, and I gave my response, which was later published with full statistical calculations in the *New Scientist*, which had given Steele a great deal of prominence and in which he had by then published his "re-interpretation" of our data. The whole episode should be a warning to young scientists that it is dangerous to be too closely wedded to one's own hypothesis, however attractive it may be.

The second bizarre moment (as well as embarrassing *and* painful in a non-physical sense) came when I was President of the Transplantation Society at the Rome Congress in 1978. I had been fêted and treated most generously by the local organizer and his committee, as I imagine all Presidents are. However, with the congress hardly in its stride, literally from one moment to the next, I found myself "sent to Coventry" – totally ignored, cold-shouldered and isolated; I suddendly seemed no longer to exist. This moment happened to coincide with the unofficial declaration of the Society's postal ballot for the election of officers, according to which the local organizer had failed in his bid for the Vice-Presidency. Had I been expected to see to it that the local man was elected to high office? I have never managed to invent a more rational explanation, nor have I been furnished with one.

The third bizarre moment was light-hearted. In 1983 I was in Kyoto attending the International Congress of Immunology. One evening Guy Voisin and his wife Janine and I decided to have a traditional Japanese meal, and Guy had been told of a good place to go to. None of us spoke Japanese, and so the hotel porter had given the taxi driver careful instructions on how to get to the restaurant. We finished up in a very dark street outside a large open gate, to be welcomed by some ladies in resplendent kimonos. One of them, apparently a little distraught, escorted us up a long and winding hillside, an elegant driveway of exquisitely raked gravel and lined by flaming torches. When we arrived at the top we were told by another very agitated lady, again dressed in a gorgeous kimono, that the restaurant had already closed. After some vexed attempts to clarify the situation we noticed that there were several pairs of men's shoes on the verandah, and in due course several Japanese men emerged and were ceremoniously seen off. We then realized that there must have been a parlous misunderstanding between the doorman and Guy, and that this "restaurant" was unlikely to provide us with the sustenance that we had anticipated. So we beat a hasty retreat in as dignified a manner as we could in the circumstances and eventually found ourselves a small but excellent place to sample the delights of Japanese cuisine.

The most disappointing moment was when on the day before Christmas 1988 my haematological and obstetric collaborators (David Linch and Charles Rodeck) and I had to meet to decide, with great reluctance, to more or less terminate a research project that had been designed to show that it is possible to induce tolerance and create red cell chimerism in primate foetuses without graft-versus-host disease. Our previous work in mice and on human foetal lymphocytes had suggested that this could become a practicable form of treatment for foetuses shown to be suffering from life-threatening bone marrow defects. Unfortunately, the technical problems in M. fascicularis and the vast expense of maintaining the monkeys, coupled with some dispiriting results which were presented to a recent meeting of the British Transplantation Society [10], proved too much. A great pity, for we thought that it should be possible to apply the concept of immunological tolerance to the treatment of congenitally defective foetuses in a direct way. We still think that the notion is sound, and I have no doubt at all that somebody else will in due course show that it can be done.

Enough of these rather disjointed reflections. I shall shortly be retiring. I used to think that retirement would be the end of the world for me, but I am beginning to think that there might indeed be life after that kind of death, and that there might even be some benefits. I look back on four decades of research with a mixture of much satisfaction and some regrets. I think I can say that most of this time has been enjoyably, if not always profitably, spent. I wish all the younger members of the Kitzbühel/Axams fraternity, and indeed immunologists throughout the world, every possible success in the quest for solutions to the many outstanding problems that my generation has bequeathed to them.

References

1. Medawar PB (1986) Memoirs of a Thinking Radish. Oxford University Press
2. Billingham RE, Brent L (1957) A simple method for inducing tolerance of skin homografts in mice. Transplant Bull 4:67–71
3. Brent L, Berrian JH (1958) Cell-bound antibodies in transplantation immunity. Ann NY Acad Sci 73:654–662
4. Brent L (1958) Tissue transplantation immunity. Prog Allergy 5:271–348
5. Brent L, Brown JB, Medawar PB (1962) Quantitative studies on tissue transplantation immunity. VI. Hypersensitivity reactions associated with the rejection of homografts. Proc R Soc Lond, Ser B 156:187–209
6. Kilshaw PJ, Brent L, Pinto M (1975) Suppressor T cells in mice made unresponsive to skin allografts. Nature 255:489–491
7. Gorczynski RM, Steele EJ (1980) Inheritance of acquired immunological tolerance to foreign histocompatibility antigens in mice. Proc Natl Acad Sci USA 77:2871–2875
8. Brent L, Rayfield LS, Chandler P, Fierz W, Medawar PB, Simpson E (1981) Supposed lamarckian inheritance of immunological tolerance. Nature 290:508–512
9. Brent L, Chandler P, Fierz W, Medawar PB, Rayfield LS, Simpson E (1982) Further studies on supposed lamarckian inheritance of immunological tolerance. Nature 295: 242–244
10. Brent L, Linch DC, Rodeck CH, Nicolaides K, DeGiorgi L, Sherwood RA, Al-Imara L, Mew J, Socha W (1989) On the feasibility of inducing tolerance in man: a study in the cynomolgus monkey. Immunol Lett 21:55–62

Problems Relating to the Transplantation of Discordant Xenografts*

F.H. Bach

Introduction

It is an enormous pleasure to participate in a meeting honoring that most remarkable man, academician, and leader in medical/scientific research, Prof. Walter Brendel. Brendel is very much respected by the community of medical scholars for the important and, in cases, dominant role that he played in various areas of biomedical investigation. For some of us who were fortunate enough to know him personally, he represented the ideal of colleague, friend, and savant. The accomplishments that surround Walter are well known and very great indeed; his particular contribution made through the organization of the Kitzbühel-Axams meetings is well-nigh unique. Walter was able to assemble an unusual group of medical investigators, most of whom were outstanding in their fields, who made attendance at the Brendel meetings their absolute first priority. Those of us who had the opportunity to participate in those meetings realize that Walter created for us a marvelous atmosphere in which to exchange ideas, establish working collaborations (each of us can probably trace some of the more important contributions from our laboratories to these meetings), and most of all establish some of the warmest, most rewarding, and lasting friendships of our lives.

It is thus with the greatest of respect that I offer this contribution in honor of Professor Brendel. I cannot do this, however, without acknowledging the very important roles in all of this that Profs. Konrad Messmer and Claus Hammer have played over the years. These individuals contributed so much to the stimulating atmosphere as well as participating so actively and productively in the scientific aspects of the meetings and discussions. Lastly, who of all of us could have managed without the gracious helpfulness extended by Mechthild Stein. To her I offer my gratitude for so many things and especially for the friendship that she has allowed.

While my contributions to this meeting over the years have, for the most part, dealt with the genetics and biology of the HLA complex and with T-lymphocyte biology, I have chosen as the subject of this article a topic that I addressed at the meetings much less frequently. Certainly one of the greatest contributions made by the Brendel team starting in 1971 was in the area of

* This is paper #548 from the Immunobiology Research Center.

K. Messmer M. Stein (Eds.)
Pathways in Applied Immunology

xenotransplantation. I do not think it hyperbole to say that this group for many years provided the driving force for the entire medical community in experimental studies of xenotransplantation. Since this is a topic in which I and several of my colleagues have again taken a very active interest, I would like to devote this article to xenotransplantation[1].

Transplantation of kidneys, hearts, livers, pancreata, and other organs has enjoyed extraordinary and increasing success in recent decades. This accomplishment has generated a problem of its own: donor organ shortages have placed clear limits on the number of patients that can be treated. Even optimistic estimates forecast a continued scarcity of certain donor organs. For this reason, the transplantation community has turned, once again, to the possibility of using xenogeneic organs for transplantation to man, an area which Brendel and colleagues have pioneered [2]. Attention has turned to the use of pigs as donors, given the acceptable size, availability, and potential for genetic manipulation of these animals. It is my purpose in this article[1] to review the particular problems that transplantation of vascularized organs from pig to man would pose and to discuss our present investigations in these areas.

It has been known since early in this century that when an organ from an animal of one species is anastamosed to an animal of a phylogenetically distant species, the organ rapidly and inexorably fails due to hyperacute rejection [3–6]. Sir Roy Calne, an active participant in these meetings, observed that recipients of such grafts have circulating antibodies that can agglutinate red blood cells (RBC) of the donor species; such species combinations were defined as discordant [7]. The antibodies, which exist in all members of a species without overt evidence of sensitization, were called natural antibodies [8]. Closely related species such as humans and chimpanzees have little or no detectable antibody that delineates phylogenetic boundaries, and such combinations are defined as concordant [7].

Depending on the species combination involved [2,6], hyperacute rejection of a vascularized xenograft can occur in as little as 10 min or in a very few hours, in contrast to cell-mediated allograft rejection which takes place in 7–10 days. While a cellular immune response would likely play a role in rejection of a discordant, immediately vascularized xenograft that survives for a sufficient number of days [9], it is hyperacute rejection that must be understood and circumvented if clinical xenotransplantation is to become a reality.

---

[1] This article is in large measure taken from a recently published review from the Minnesota Xenotransplantation Program, of which the author is the program director [1]. This group includes as principal investigators: Drs. Bolman, Dalmasso, Matas, Najarian, Platt and Vercellotti, in addition to the author.

A Model of Hyperacute Rejection

We hypothesize that natural antibodies of the recipient bind to endothelial cells lining the vessels of the donor organ, the complement (C) cascade is triggered, and the endothelial cells are as a result "activated." The concept of endothelial cell activation is taken from recent studies which have portrayed a variety of metabolic changes that occur when endothelial cells are perturbed by noxious stimuli such as tumor necrosis factor (TNF) or interleukin 1 (IL-1). These changes include, among others, the synthesis of tissue factor, loss of cell surface thrombomodulin, the release of platelet-activating factor which stimulates the aggregation and adhesion of platelets, expression of cell surface molecules which promote adhesion of inflammatory cells, and disruption of the barrier to efflux of plasma proteins and blood-borne cells [10–16]. Endothelial cell activation may thus result in interstitial hemorrhage and inflammation, edema, and small vessel thrombosis, which together comprise the pathology of hyperacute rejection.

The information available regarding mechanisms of hyperacute rejection comes from several lines of investigation. I shall discuss primarily two in this article. First, I consider findings from immunopathological examination of grafts that are undergoing various stages of hyperacute rejection. Secondly, we have used an in vitro model of hyperacute rejection in which we study the effects of human natural antibodies and complement (in the form of human or rhesus serum) on cultured porcine endothelial cells. We have, as have others [17,18], studied the consequences of depleting natural antibodies from potential recipients and assaying the effects of such depletion on graft survival.

Immunopathology of Hyperacute Rejection

Natural Antibodies. Immunofluorescence photomicrographs of sections of a pig heart that had been transplanted to a rhesus monkey recipient and examined 2 h later shows the deposition of IgM along the endothelium lining of donor organ vessels. IgG is not deposited on endothelial cells but is present in the interstitium in the same distribution as albumin, suggesting that the presence of IgG is secondary to nonspecific diffusion. We interpret these findings as showing that in the pig-to-rhesus combination, natural antibodies in a nonsensitized recipient are primarily, if not solely, IgM.

Complement. There is codeposition of C3 and the membrane attack complex with IgM in hyperacutely rejecting grafts. C4, an integral component of the classical pathway, is similarly fixed whereas properdin and factor B of the alternative pathway are absent. These findings suggest that it is the classical, and not the alternative, pathway of C that is involved in hyperacute rejection in this species combination. Prolonged survival seen after administration of cobra venom factor (an inhibitor of C) is also consistent with an integral role of C in the pathogenesis of hyperacute rejection [19–21].

Other Factors. In addition to natural antibodies and C, and likely consequent to their deposition on endothelial cells, platelets, neutrophils, and fibrin are deposited along the endothelium. There is evidence that these factors, together with vasoconstriction [6,22–25] play an important role in intravascular coagulation and tissue ischemia that are the hallmarks of hyperacute rejection.

The model presented above suggests that it is the combined action of natural antibodies and C on endothelial cells that initiates the events resulting in hyperacute rejection. I discuss below the results of in vitro studies aimed at defining the mechanisms through which hyperacute rejection takes place.

Natural Antibodies and Their Targets

Xenoreactive natural antibodies have classically been studied based on hemagglutination or lymphocytotoxicity [2,6–8,26]. While these studies have produced valuable data, it has been uncertain to what extent RBC or lymphocytes express on their surfaces the same antigens, in the same proportionate representation as are expressed on endothelial cells. Because of this uncertainty we have developed an enzyme-linked immunosorbent assay for natural antibodies using cultured porcine endothelial cells as the target cells [27]. Indeed in subsequent studies we found (as described below) that the molecules recognized by human natural antibodies on pig endothelial cells are in part different from those recognized on pig RBC and lymphocytes.

The definition of the antigens, and their determinants, that are recognized by natural antibodies is a critical issue about which very little is known; acquisition of such knowledge could lead to specific extracorporeal immunoabsorption of natural antibodies on columns to which those target determinants are bound, as well as other manipulations that may have therapeutic potential. Previously published data [2,28] suggested, based upon immunoabsorption and blocking studies, that the determinants (or cross-reactive structures) recognized by natural antibodies on endothelial cells might also be represented on hemopoietic cells, and that the molecules recognized vary between species. The studies, however, provided little insight into the biochemical nature of the target antigens.

We have extended these studies in an attempt to define further the molecules and determinants recognized by human and rhesus natural antibodies on cultured porcine endothelial cells [29]. Our results indicate that rhesus and human natural antibodies predominantly recognize a triad of glycoproteins, which we call gp115/135, but not glycolipids, phospholipids, or proteoglycans that are expressed on the surface of porcine aortic endothelial cells. The major determinants appear to be located on oligosaccharide substitutions associated with these molecules since the binding of natural antibodies in solid-phase assay is eliminated by treatment of the glycoproteins with various glycosidases but not by treatment with proteases. The biological importance of these glycoproteins and several moieties of lower molecular weight is suggested by the observations that (a) perfusion of porcine kidney by rhesus blood absorbs natural antibodies against these antigens and enables prolonged xenoengraftment of a porcine heart transplanted into the treated animal and (b) rejection of that heart is temporally associated with return to the circulation of antibodies against those moieties.

Activation of Endothelial Cells

Endothelial cells function not only to form a physical barrier through which the blood and its elements can flow and across which gases and fluids can transverse but also to inhibit the formation of thrombi which might impair the flow of blood. When endothelial cells are activated, however, the sequelae lead to a loss of vascular integrity and the development of a procoagulant environment around the endothelial cells.

The in vitro system that we have used as a model of hyperacute rejection consists of cultured pig aortic endothelial cells and studies the consequence of adding human natural antibodies and C (in the form of human serum) to these cells. We have evaluated whether natural antibodies and C activate endothelial cells in a manner similar to that observed when endothelial cells are activated by agents such as endotoxin, IL-1, or TNF [10–15]. Our initial investigations suggested that the binding of natural antibodies to endothelial cells and the activation of complement was associated with increased polymorphonuclear neutrophil (PMN) adhesion mediated at least in part by complement receptor type 3 on PMN and a rapid perturbation of the cells as measured by enhanced Ca^{2+} flux. Subsequently it was shown that endothelial cells synthesized tissue factor and lost thrombomodulin. In addition, we have investigated whether endothelial cell activation leads to hitherto unknown phenomena associated with potential loss of vascular integrity and procoagulation.

Release of Heparan Sulfate Proteoglycan from Activated Endothelial Cells

Our studies have defined a previously unrecognized consequence of activating endothelial cells with natural antibodies and C: heparan sulfate proteoglycan is lost from the surface of the endothelial cells [30,31]. Heparan sulfate proteoglycan in the cell membrane and extracellular matrix of normal blood vessels is thought to contribute to three of the physiological properties of blood vessels: (a) maintenance of an antithrombotic cell surface [32], (b) anchoring of superoxide dismutase which prevents oxidant injury [33], and (c) maintenance of a barrier to the egress of plasma proteins and blood cells [34,35]. Since failure of these processes might well contribute to the pathological picture of hyperacute rejection, we tested whether binding of natural antibodies to endothelial cells and the activation of C might affect heparan sulfate proteoglycan. Significant release is seen within 4 min, and greater than 50% release is present by 60 min. This represents endothelial cell activation and not impending death, since even at 60 min there is no morphological evidence of irreversible damage to the endothelial cells, and the cells still incorporate amounts of radioactive uridine and thymidine similar to those incorporated by controls.

Release of heparan sulfate proteoglycan is dependent on the activation of classical pathway of C; the alternative pathway of C does not appear to be involved [31]. Significant levels of heparan sulfate proteoglycan release mediated by sera deficient in the later components of C, at least suggests that the entire C cascade may not be essential for endothelial cell activation and release of heparan sulfate proteoglycan.

Cellular Immunity in Widely Discordant Xenogeneic Combinations

It has not been possible to study cellular immunity to antigens on cells of immediately vascularized discordant xenografts given the explosive nature of natural antibody-mediated hyperacute rejection; the cell-mediated immune reaction in such situations does not have time to develop before the graft is rejected. Studies of cellular immune responses to discordant grafts have thus been restricted to (a) in vitro investigations and (b) studies of cellular response to neovascularized tissues such as pancreatic islets or skin.

In vitro studies demonstrated, in concert with findings of Simonsen [36] using the chorioallantoic membrane of the chick as a target organ, that xenogeneic proliferative responses in mixed leukocyte culture (MLC) do occur but are weaker than allogeneic responses [37,38], apparently the more so in relation to the incremental disparity between the two species being tested. This attenuated response was complemented by the observation that in species combinations such as man–anti-mouse, or the opposite direction, the response was dependent on the presence in the MLC of responder antigen-presenting cells (APC) [39], unlike an allogeneic MLC, in which either responder or stimulator APC would suffice to support a response [40]. An important question is thus whether the ability to recognize highly disparate antigens (e.g., human recognizing mouse) is included in the human T cell repertoire. Our recent data show that the human T cell repertoire does include the ability to recognize murine xenoantigens, but that the development of a proliferative response requires that human cytokines be added to the culture. This observation suggests that the cytokines produced by murine APC do not adequately subserve the response of human T cells to mouse antigens [41]. This, however, does not appear to be the whole story. Sherman and colleagues [42] have recently studied the ability of murine T cytotoxic cells to recognize human class I antigens. In this case, response could be obtained if the murine class I α-3 domain was inserted into the human class I antigen, suggesting that corecognition of the target class I molecule by CD8 was essential to allow the direct response to the human antigen.

We favor the following model for response to xenoantigens in the human-mouse combination. We believe that in both directions of response, there is a need for T cell receptor (TcR) recognition of xenogeneic MHC presented either directly or perhaps more likely indirectly in association with MHC class II. In addition, there exists a need for corecognition by CD4 or CD8 of the xenogeneic class II or class I antigen, respectively, plus the availability of the appropriate cytokines. Thus, the human anti-mouse response would likely depend on the binding of human TcR to specific mouse MHC sequences and the corecognition of the murine MHC by human CD4. In this case the relative inability of murine IL-1 to provide a signal to the human T cells, would require as well the overt addition of the human IL-1. In the mouse anti-human reaction, there is the need to substitute the murine α domain to allow CD8 corecognition of human MHC but no corresponding need for murine cytokines as the cytokines produced by the human T cells can be used by murine T cells.

The argument has recently been made that in widely disparate, discordant xenografts it is $CD4^+$ cells that play the major role in graft rejection, having an even greater relative role than in allograft rejection [43]. To what degree both $CD4^+$ cells and $CD8^+$ cells are functional in leading to rejection needs further evaluation. These results have been obtained, for reasons discussed earlier, in neovascularized grafts. Whether a similar picture would be obtained in the immediately vascularized grafts is also not known.

Clinical Xenotransplantation: Achieving Accommodation

One approach to successful xenotransplantation would be to aim for the elimination of natural antibodies from the circulation by immunoabsorption [18] or by depletion of the B cells that synthesize such antibodies. Natural antibodies present in a rhesus recipient are extensively depleted after absorption by porcine kidney perfusion and after xenoengraftment of a porcine heart. Kidneys used for immunoabsorption are hyperacutely rejected. The immunopathology of the rejected heart, which can survive for 3–5 days after such natural antibody depletion, is quite similar to that observed in the hyperacute rejection that occurs within a few hours in unmodified recipients. The time of rejection of the hearts that we have studied is consistent with the extensive experience reported by Cooper et al. [18].

It is likely that coincident with the depletion of natural antibodies by organ perfusion there is depletion of C and possibly other factors that are deposited in the kidneys. How much depletion of factors other than natural antibodies contributes to the prolonged survival of the heart is not known although our limited experience indicates that complement levels may return to the normal range for a period of days before rejection occurs and thus suggests that C does not itself (through alternative pathway activation) incite the rejection reaction. Prolongation of heart survival can also be obtained by removal of natural antibodies by repeated plasmaphereses [17], in which case several factors in addition to natural antibodies and complement are also depleted.

We believe that the success of allotransplantation in two different situations in which there initially exist in the recipient antibodies directed at donor tissues may provide an alternative model for an approach to clinical xenotransplantation. In cases of ABO incompatibility, in which anti-A1 antibodies are present in the recipient, transplantation of kidneys results in hyperacute or accelerated rejection in a very great majority of cases. Several different centers, including our own, have shown that depletion of those antibodies before transplantation, with maintenance of low antibody levels for an ill-defined period (perhaps around 2 weeks) after transplantation, leads to a state in which the graft survives even though the antibodies later return. In the ABO system, we have shown that the antibodies that return are directed at the ABO antigens of the donor, and that those antigens endure on the endothelial cells of the surviving graft [44]. Similar findings, although less extensively studied, have been observed following removal of anti-HLA antibodies [45]. We have used the term "accommodation" to refer to this phenomenon to reflect the hypothesis which is the basis of our preclinical efforts.

There are several possible explanations for accommodation. First, for a variety of reasons, the susceptibility of endothelial cells of the graft to injury may vary over time. As a consequence of the transplant procedure or exposure to preservation solution endothelial cells may become especially sensitive to injury caused by the action of natural antibodies and C. At a later time, possibly days or weeks after engraftment, endothelial cells may achieve a condition in which they resist such injury as might be elicited by the return of natural antibodies to the circulation. Further, endothelial cells may, as a consequence of exposure to gradually increasing concentrations of natural antibodies change in such a way as to increase the resistance to injury.

Secondly, it is possible that after depletion of natural antibodies and the presence of the graft for some period of time, the antibodies that return to the circulation differ in affinity and/or specificity from the natural antibodies present before grafting. We have noted that following the transplantation of a pig heart to a rhesus monkey from which natural antibodies had been depleted, the antiendothelial antibodies present on day 4, (12 h after graft rejection), included IgM and IgG isotypes. Whether the IgG antibodies are products of an Ig class switch in the B cells producing the natural antibodies is not clear. A change in antibody specificity might result in a decrease in antibody binding to the graft and a change in subclass, leading to a decrease in complement activation.

Thirdly, it may be that the epitopes normally expressed by endothelial cells are changed or modulated after the transplantation procedure. We have observed in two recipients of ABO-incompatible kidney grafts that after anti-A or anti-B antibodies returned to the circulation, there was no evidence of antibody deposition in the graft. While the endothelium of the graft continued to express A or B antigen, it is possible that core structures were changed in such a way to decrease interaction with circulating antibody.

We have recently studied one monkey in which repeated plasmaphereses and administration of immunosuppressive agents were performed, and 2 weeks later sequential porcine kidneys were perfused, again under cover of immunosuppression followed by transplantation of a porcine heart. The recipient animal had to be sacrificed on the 8th postoperative day, at which time the heart was still beating and evidenced normal ECG activity. Most interesting was that on immunopathological examination on day 8, rhesus IgM was deposited on the endothelium of the heart although there were only focal, and very limited, deposits of C and fibrin, a pathological picture very similar to that seen with one of the cases of accommodation in ABO incompatibility that we have studied. While these findings have to date been observed in only a single animal, the observations encourage the approach of attempting to achieve accommodation, as occurred in this recipient, as a possible solution to the problem of discordant xenografting. Also encouraging is the work of Alexander and colleagues who found in a minipig-to-baboon model that the temporary removal of natural antibodies by plasmapheresis enabled prolonged (up to 3 weeks) engraftment of a xenogeneic kidney [17]. It is uncertain whether this relative success represents accommodation as described above.

Clearly this is an area that will require much additional experimentation, probably including the testing of therapeutic agents that interfere with certain of the endothelial cell-related phenomena discussed above. Hopefully, given the clinical imperative that exists, many more investigators will enter this area, which has been so fruitfully investigated by Brendel, Hammer, and colleagues, and will contribute to what is both a challenging and potentially clinically rewarding quest.

Acknowledgements. The author thanks all members of the xenotransplantation group at the University of Minnesota and especially Dr. Jeffrey Platt, with whom all this work was discussed and for much of which he deserves major credit.

References

1. Platt J, Vercellotti G, Dalmasso A, Matas A, Bolman R, Najarian J, Bach FH (1990) Transplantation of discordant xenografts: a review of progress. Immunol Today, Vol. 11, No. 12:450–457
2. Hammer C (1987) Isohemagglutinins and preformed natural antibodies in xenogeneic organ transplantation. Transplant Proc 19:4443–4447
3. Reemtsma K, (1989) Xenotransplantation: a personal history. In: Hardy MA (ed) Xenograft 25. Elsevier, Amsterdam, pp 7–16
4. Perper RJ, Najarian JS (1966) Experimental renal heterotransplantation. I. In widely divergent species. Transplantation 4:377–388
5. Giles GR, Boehmig HJ, Lilly J et al. (1970) Mechanism and modification of rejection of heterografts between divergent species. Transplant Proc 2:522–537
6. Auchincloss H (1988) Xenogeneic transplantation. Transplantation 46:1–20
7. Calne RY (1970) Organ transplantation between widely disparate species. Transplant Proc 2:550–556
8. Boyden SV (1964) Natural antibodies and the immune response. Adv Immunol 5:1–28
9. Perper RJ, Najarian JS (1966) Experimental renal heterotransplantation. II. Closely related species. Transplantation 4:700–711
10. Pober JS (1988) Cytokine-mediated activation of vascular endothelium. Am J Pathol 133:426–433

11. Bevilacqua MP, Pober JS, Mendrick DL (1987) Indentification of an inducible endothelial-leukocyte adhesion molecule. Proc Natl Acad Sci USA 84:9238–9242
12. Nawroth PP, Stern DM (1986) Modulation of endothelial cell hemostatic properties by tumor necrosis factor. J Exp Med 163:740–745
13. Brett J, Gerlach H, Nawroth P et al. (1989) Tumor necrosis factor/cachectin increases permeability of endothelial cell monolayers by a mechanism involving regulatory G proteins. J Exp Med 169:1977–1991
14. Sherry B, Cerami A (1988) Cachectin/tumor necroisis factor exerts endocrine, paracrine, and autocrine control of inflammatory responses. J Cell Biol 107:1269
15. Schleaf RR, Bevilacqua MP, Sawdey M et al. (1988) Cytokine activation of vascular endothelium. Effects on tissue-type plasminogen activator and type I plasminogen activator inhibitor J Biol Chem 263:5797–5803
16. Ryan US (1986) The endothelial surface and responses to injury Fed Proc 45:101–108
17. Alexandre GPJ, Gianello P, Latinne D et al. (1989) Plasmapheresis and splenectomy in experimental renal xenotransplantation. In: Hardy MA (ed) Xenograft 25. Elsevier, Amsterdam, pp 259–266
18. Cooper DKC, Human PA, Lexer G et al. (1988) Effects of cyclosporine and antibody adsorption on pig cardiac xenograft survival in the baboon. J Heart Transplant 7: 238–2246
19. Miyagawa S, Hirose H, Shirakura R et al. (1988) The mechanism of discordant xenograft rejection. Transplantation 46:825–830
20. Gewurz H, Clark DS, Cooper MD et al. (1967) Effect of cobra venom-induced inhibition of complement activity on allograft and xenograft rejection reactions. Transplantation 5:1296–1303
21. Mejia-Lagune JE, Martinez-Palomo A, Biro CE et al. (1972) Morphologic study of the participation of the complement system in hyperacute rejection. Am J Pathol 69:71–78
22. Marceau JP, Hallenbeck GA, Zollman PE et al. (1965) A comparison of autoplastic, allogeneic and xenogeneic perfusion of isolated kidneys. J Surg Res 5:492–502
23. Rosenberg JC, Hawkins E, Rector F (1971) Mechanisms of immunological injury during antibody-mediated hyperacute rejection of renal heterografts. Transplantation 11: 151–157
24. Busch GJ, Martins ACP, Hollenberg NK et al. (1975) A primate model of hyperacute renal allograft rejection. Am J Pathol 79:31–57
25. Busch GJ, Martins ACP, Hollenberg NK et al. (1976) Successful short-term modification of hyperacute renal allograft rejection in the primate. Am J Pathol 82:43–60
26. Hammer C (1989) Preformed natural antibodies (PNAB) and possibilities of modulation of hyperacute xenogeneic rejection (HXAR). Transplant Proc 21:522–523
27. Platt JL, Turman MA, Noreen H et al. (1990) An ELISA assay for xenoreactive natural antibodies. Transplantation 49:1000
28. Cameron DJ, Rajagopalan, Fitts CT, Majeski JA (1983) Characterization of the preformed antibodies involved in the xenograft rejection. J Surg Oncol 22:157–163
29. Platt JL, Lindman BJ, Chen H, Spitalnik SL, Bach FH (1990) Endothelial cell antigens recognized by xenoreactive human natural antibodies. Transplatation, Vol, 50, 5: 817–822
30. Platt JL, Dalmasso AP, Vercellotti GM (1990) Endothelial cell proteoglycans in xenotransplantation. Transplant Proc 22:1066
31. Platt JL, Vercellotti GM, Lindman B (1990) Release of heparan sulfate from endothelial cells: implications for pathogenesis of hyperacute rejection. J Exp Med 171:1363–1368
32. Marcum JA, Rosenberg RD (1987) Anticoagulantly active heparan sulfate proteoglycan and the vascular endothelium. Semin Thromb Hemost 13:464–474
33. Karlsson K, Marklund SL (1988) Plasma clearance of human extracellular superoxide dismutase C in rabbits. J Clin Invest 82:762–766
34. Gallagher JT, Lyon M, Steward WP (1986) Structure and function of heparan sulphate proteoglycans. Biochem J 236:313–325

35. Matener Y, Bar-Ner M, Yahalom J et al. (1985) Degradation of heparan sulfate in the subendothelial extracellular matrix by a readily released heparanase from human neutrophils. Possible role in invasion through basement membranes. J Clin Invest 76: 1306–1313
36. Simonsen M (1967) The clonal selection hypothesis evaluated by grafted cells reacting against their hosts. Cold Spring Harb Quant Biol 32:517–523
37. Wilson DB (1967) Quantitative studies on the mixed lymphocyte interaction in rats. J Exp Med 126:655–665
38. Lindahl KF, Bach FH (1975) Human lymphocytes recognize mouse alloantigens. Nature 254:607–609
39. Lindahl KF, Bach FH (1976) Genetic and cellular aspects of xenogeneic mixed leukocyte culture reaction. J Exp Med 144:305–318
40. Alter BJ, Bach FH (1990) Cellular basis of the proliferative response of human T cells to mouse xenoantigens. Cell Immunol 1:207–218
41. Alter BJ, Bach FH (1990) Allelic T-cell receptor alpha complexes have little or no infuence on susceptibility to type 1 diabetes. J Exp Med 71:333–338
42. Irwin MJ, Heath WR, Sherman LA (1989) Species-restricted interactions between CD8 and the a3 domain of class I influence the magnitude of the xenogeneic response. J Exp Med 170:1091–1101
43. Pierson RN, Winn HJ, Russel PS et al. (1989) Xenogeneic skin graft rejection is especially dependent on $CD4^+$ T cells. J Exp Med 170:991–996
44. Chopek M, Simmons R, Platt JL (1978) ABO incompatible renal transplantation: initial immunopathologic evalution. Transplant Proc 19:4553–4557
45. Palmer A, Welsh K, Gjorstrup P et al. (1989) Removal of anti-HLA antibodies by extracorporeal immunoadsorption to enable renal transplantation. Lancet i:10–12

Minor Histocompatibility Antigens

Elizabeth Simpson

Introduction

It is appropriate at this time that we should be reflecting on the past as well as looking into the future. During this series of 20 meetings on applied immunology organised by Walter Brendel, I have enjoyed spending time with a number of people who have influenced me in science. Peter Medawar was a very strong influence on my career, and he introduced me to a large number of people, including some in this audience, and of course Walter Brendel. Peter and Walter shared, amongst other things, a terrific desire to see basic science applied to medicine. They have both in their own very different ways been extraordinarily successful in encouraging others to realise the importance of applying basic science to solving clinical problems. I thank Walter for his part in making this apparent not only to me but to a wider audience and for fostering, amongst people of different backgrounds and ages in these meetings, friendships and collaborations which have been very much in the Medawar tradition. I thank him and all his colleagues who have made these meetings possible and have enabled me to take part in the inherent enjoyment and instruction of them.

Minor Histocompatibility Antigens Defined

Minor histocompatibility (H) antigens were first defined by Snell [1] and Bailey [2] as those genetic differences between mice matched at H-2, the mouse major histocompatibility complex (MHC) [3], giving rise to rejection of organs or tissues transplanted between them. Genetically, the loci encoding such minor H antigens [2] map outside the MHC, H-2, which is on chromosome 17 in mice [3]. By making appropriate genetic crosses and congenic mouse strains, Snell and Bailey were able to localise the genes encoding the minor H antigens into linkage groups and then onto chromosomes. Minor H loci in mouse occur on almost every chromosome and number at least 40; the same is likely to be true in man, in whom immune responses to minor H antigens are also manifest as host-versus-graft and/or graft-versus-host responses in recipients transplanted with organs or tissues from MHC (HLA) matched donors. Graft-versus-host disease remains a particular problem in bone marrow transplantation [4].

K. Messmer M. Stein (Eds.)
Pathways in Applied Immunology

In Vitro Responses to Minor H Antigens

In vivo responses to H antigens indicate the immune response capacity of the whole animal, but our ability to investigate these responses further has been very greatly helped by the development of in vitro responses to these antigens. This has enabled us to examine not only the classes of lymphocytes involved in responding but also the nature of the antigen exciting the response. Specific immune responses can in general be made by T and B lymphocytes [5]. In the case of MHC (H-2 and HLA) allo-antigens, T cells of the two major subclasses, Tc (focussed on MHC class I molecules) and Th (focussed on MHC class II molecules) respond and can be cultured in vitro. B cells also respond by making antibodies specific for allogeneic class I or class II molecules; all the initial analysis of MHC antigens was made using these antibodies, allowing a molecular characterisation of them and the cloning of genes encoding them.

In contrast, minor H antigens, like many tumour-specific antigens, elicit only T cell responses, and whilst minor H specific T cell clones can be grown continuously in vitro and used to type cells from humans or mice for the expression of those antigens [6], T cell clones are more difficult to use than antibodies. This is because of the rules of MHC restriction (the need to match MHC antigens on the cells to be typed [7]), and it is certainly not so straightfoward to use T cell clones as antibodies for molecular characterisation of either minor H antigens or the genes that encode them.

Peter Medawar was an enthusiastic supporter of our work using T cell responses to analyse minor H antigens and immune responses to them. This was in large part because he saw the likeness between minor H antigens and tumour-specific antigens, which also often behaved as weak transplantation antigens. So not only were minor H antigens of interest to him as having a role in transplantation, the study which he pioneered, but he saw them also as a paradigm for tumour-specific antigens, which excited his interest during the years that we spent working at the Clinical Research Centre in Harrow. It was Peter who introduced me during this time to Thierry Boon in whose laboratory in Brussels the first tumour-specific antigen gene was cloned by an ingenious use of T cell clones specific for tumour-specific antigens [8,9]. Paradoxically, molecular advances in this area are showing us ways of tackling questions of the molecular nature of minor H antigens.

MHC Restriction, Allorecognition and Endogenous Peptides

The crystal structure of the human HLA-A2 molecule was solved by Bjorkman in 1987 [10]. The paper in which the first pictures of this structure appeared was published the week that Peter Medawar died, and I took that copy of *Nature* to read in the train on my way to his funeral. I felt saddened that he never saw those pictures – they would have delighted him. They gave us a profound yet simple explanation for the phenomenon of MHC restriction, discovered more than 10 years previously [7], and which established the need for T cells to use self MHC antigens as guidance molecules. The specificity of T cells always involves not only the nominal antigen but also a self MHC molecule. This is as true for viral antigens and exogenous protein antigens as well as for minor H antigens [5]. Now we could see why: the MHC molecule contains an antigen (peptide-sized) binding groove, bounded on each side by two α-helices and at its base by eight β-pleated sheets. The epitope of antigen plus MHC molecule recognised by T cell receptors consists of the peptide and the adjacent parts of the antigen-binding groove. All the polymorphic residues of an MHC molecule lie within the α-helices or the β-pleated sheets at sites where they form the lining of the antigen-binding groove [11]. Here they determine differences in the ability to bind different peptides from exogenous or endogenous antigens and hence can account for the immune response gene function of MHC genes. Individuals who possess alleles of particular MHC molecules which provide a better fit for certain peptides can thereby make a better immune response than individuals whose MHC antigen-binding grooves give a poorer fit. The structure of MHC class II molecules has now been modelled on the solved structure of class I and provides a perfect analogy for class II restricted responses.

It is very likely that minor H antigens, which elicit both class I restricted responses by $CD8^+$ T cells and class II restricted responses by $CD4^+$ T cells are peptide products of endogenous self molecules: alleles of minors could constitute peptides differing at perhaps as little as a single amino acid. Their function as transplantation antigens would be a by-product of such substitutions. The physiological function of the endogenous molecules which give rise to them could be various and unrelated and can be discovered only when the peptides and their origins are identified. However, one consequence of the existence of allelic forms of peptides from endogenous molecules within the peptide-binding grooves of self MHC antigens is that there they can exert an influence on the T cell repertoire, and there is very good evidence from experiments in mice that minor H antigens provide very powerful selective forces in shaping the T cell repertoire, both by negative and positive selection of T cells expressing particular V region genes (Vβ and Vα) [12–14]. In this way, they might account for some of the non-MHC immune response (Ir) genes known to map to a number of different chromosomes [15]. Examples of endogenous self molecules which can select the T cell repertoire are Mls^a, Mls^c, the B cell specific self antigen recognised in association with H-2E molecules in mouse strains able to use Vβ17a for T cell receptors [16] and another tissue-specific self antigen recognised in association with H-2E molecules by T cells expressing Vβ11 in some mouse strains [17]. An example of a non-MHC Ir gene encoding a self antigen causing tolerance is seen in certain $H\text{-}2^d$ strains unable to respond to the synthetic polymer GT: in this case GT mimics a self antigen, and the strain which expresses the self antigen eliminates all T cells able to respond to it [18]. More examples of these types are likely to surface and to account for the fine tuning of the immune response to exogenous antigens, even when MHC Ir genes account for the greater part of the control, and, indeed, the non-MHC Ir genes are all likely to operate via the

MHC ones, since their products are proposed to reside in the antigen-binding grooves of MHC molecules.

The recognition of allogeneic MHC molecules by T cells has certain peculiar characteristics which need to be accounted for. The most obvious one is the extraordinarily high frequency of both class I and class II reactive T cells to allogeneic MHC antigens. At one time it was proposed that this was due to skewing of T cell receptor (TCR) germ line genes towards those with specificity for MHC molecules of the species, but the appearance of the Bjorkman crystal structure of class I molecules has made more likely a hypothesis originally put forward in 1977 by Matzinger and Bevan [19], namely that the specificity of allo-MHC reactive T cells is directed against the conformation of the MHC molecule itself associated with the products of other endogenously derived molecules. We would now see these as peptides. The frequency of T cells responsive to allo-MHC molecules could be accounted for by a very heterogeneous response, comprised of all the T cell clones which recognise an endogenous peptide bound to an allogeneic MHC molecule [20,21]. This alloreactive repertoire would be cross-reactive and the sum of that directed against self MHC plus exogenous peptides. Evidence consistent with this hypothesis comes from T cell clones specific for self MHC plus X which cross-reactively recognise either all cells carrying particular allo-MHC antigens or some but not all cells carrying these antigens. These data indicate that alloreactive and self MHC restricted T cells are derived from the same T cell pool.

Allorecognition of Exon-Shuffled MHC Class I Gene Products

In addition to the "classic" MHC class I antigens which serve as restriction molecules for recognition of nominal peptides, there are a number of other MHC class I molecules of very similar structure, including association with β_2m domains. Apart from the observation that they are expressed at lower levels than the classic class I molecules, and that they have a more restricted tissue distribution, there is very little to account for them (a) acting as weak allogeneic transplantation antigens and (b) failing to act as restriction molecules. To explore the cause of these differences, a series of transgenic mice have been made using exon-shuffled MHC class I genes by our collaborator, Dr. Andrew Mellor, at the National Institute for Medical Research, Mill Hill, and we have explored the expression and function of their encoded molecules. In each case, exons 2, 3 and 4 of the Q9 gene, encoding the α1, α2 and α3 domains, respectively, of the "non-classic" Qa2 molecule have been used, linked either to a 5′ promoter of the Q9 gene or of the "classic" class I D^b gene and either to the transmembrane and cytoplasmic encoding exons of Q9 or D^b. In each case, expression of Qa2 at fairly high levels on all lymphoid cells has been achieved, regardless of which 5′ promoter region was used, or whether a transmembrane polypeptide domain of D^b or a phosphatidylinositol membrane linkage characteristic of the Q9 gene was used. In each case skin grafts from mice expressing the Q9 transgenes were rejected by non-transgenic mice of the inbred strain CBA into which the transgenes were introduced, at a rate more characteristic of rejection of grafts disparate at classic MHC loci. From this result, our current concept of allorecognition would predict that such Qa2 molecules should serve as restriction molecules for recognition of endogenous viral or minor H antigens. This prediction has not yet been fulfilled, using conventional viral and minor H antigens.

Conclusions

Transplantation antigens were initially discovered as barriers to transplanting tissues and organs between genetically dissimilar members of the same species. The first ones to be identified at a molecular level in the 1960s were the MHC antigens, but it was another decade before their physiological function as guidance molecules for T cells was uncovered. There are a number of "non-classic" MHC class I antigens in both humans and mice, but they do not serve as guidance molecules in this way, for reasons that are still unclear. Minor H antigens have remained inaccessible at the molecular level, although in many cases the genes encoding them have been accurately chromosomally mapped. Their physiological function is also unclear, although consequent on their expression the T cell receptor repertoire undergoes selection. This is comprehensible in terms of their likely association with MHC molecules, probably as peptides in the antigen-binding grooves. Autoantigens, the targets of clinically apparent autoimmune disease, are examples of tissue-specific antigens that may be akin to minor H antigens. Similarly, tumour-specific antigens recognised by T cells are likely to be related moieties. The recent cloning of a tumour-specific antigen gene encoding a mutant peptide sequence provides evidence for this.

References

1. Snell GD (1948) Methods for the study of histocompatibility genes. J Genetics 49: 87–103
2. Bailey DW (1975) Genetics of histocompatibility in mice. I. New loci and congenic lines. Immunogenetics 2:249–256
3. Klein J (1975) Biology of the mouse histocompatibility-2 complex. Springer, Berlin Heidelberg New York
4. Goulmy E (1988) Minor histocompatibility antigens in man and their role in transplantation. In: Morris PJ, Tilney N (eds) Transplant reviews, Vol. 2. Saunders, Philadelphia, pp 29–54
5. Simpson E (1986) T and B lymphocytes: two repertoires or one? Immunol Lett 12: 185–193
6. Simpson E, Tomonari K (1989) Mapping minor H genes. Immunol Suppl 2:42–48
7. Zinkernagel RM, Doherty PC (1974) Restriction of in vitro mediated cytotoxicity in lymphocytic choriomeningitis within a syngeneic or semiallogeneic system. Nature 248: 701–702

8. Wolfel T, Van Pel A, De Plaen E, Lurquin C, Maryanski JL, Boon T (1987) Immunogenic (tum^-) variants obtained by mutagenesis of mouse mastocytoma P815. VIII. Detection of stable transfectants expressing a tum^- antigen with a cytolytic T cell stimulation assay. Immunogenetics 26:178–187
9. De Plaen E, Lurquin C, Van Pel A, Mariame B, Szikora J-P, Wolfel T, Sibille C, Chomez P, Boon T (1988) Immunogenic (tum^-) variants of mouse tumor P815: cloning of the gene of tum^- antigen P91A and identification of the tum^- mutation. Proc Natl Acad Sci USA 85:2274–2278
10. Bjorkman PJ, Saper MA, Samraoui B, Bennett WS, Strominger JL, Wiley DC (1987) Structure of the human class I histocompatibility antigen, HLA-A2. Nature 329:506–512
11. Bjorkman PJ, Saper MA, Samraoui B, Bennett WS, Strominger JL, Wiley DC (1987) The foreign antigen binding site and T cell recognition regions of class I histocompatibility antigens. Nature 329:512–518
12. Kappler JW, Staerz U, White J, Marrack PC (1988) Self tolerance eliminates T cells specific for Mls-modified products of the major histocompatibility complex. Nature 332:35–40
13. MacDonald HR, Schneider R, Lees RK, Howe RC, Acha-Orbea H, Festenstein H, Zinkernagel RM, Hengartner H (1988) T cell receptor Vβ use predicts reactivity and tolerance to Mls^a-encoded antigens. Nature 332:40–45
14. Pullen AM, Marrack P, Kappler JW (1988) The T cell repertoire is heavily influenced by tolerance to polymorphic self antigens. Nature 335:796–801
15. Fierz W, Farmer GA, Sheena JH, Simpson E (1982b) Genetic analysis of the non-H-2 linked Ir genes controlling the cytotoxic T-cell response to H-Y in $H\text{-}2^d$ mice. Immunogenetics 16:593–601
16. Marrack P, Kappler J (1988) T cells can distinguish between allogeneic major histocompatibility complex products on different cell types. Nature 332:840–843
17. Tomonari K, Lovering E (1988) T cell receptor specific monoclonal antibodies against a Vβ11 positive mouse T cell clone. Immunogenetics 28:445–451
18. Vidovic D, Matzinger P (1988) Unresponsiveness to a foreign antigen can be caused by self tolerance. Nature 336:222–225
19. Matzinger P, Bevan MJ (1977) Why do so many lymphocytes respond to the major histocompatibility antigens? Cell Immunol 29:1–5
20. Bernhard EJ, Le AT, Yannelli JR, Holterman MJ, Hogan KT, Parham P, Engelhard VH (1987) The ability of cytotoxic T cells to recognise HLA-A2 1 or HLA-B7 antigens expressed on murine cells correlates with their epitope specificity. J Immunol 139: 3614–3621
21. Lombardi G, Sidhu S, Lamb JR, Batchelor JR, Lechler R (1988) Co-recognition of endogenous peptide with HLA-DRI by alloreactive human T cell clones. J Immunol 142:753–759

Epitope Linkage in the Immune Response and Its Implications for Lymphoid Architecture and Immunological Memory

N.A. Mitchison

Introduction

It is a pleasure to preface this paper with an acknowledgment of how important to me has been this series of meetings that Walter Brendel has organized, and how grateful I am for having participated in them. This is not just for the unique mixture of the intellectual life with joy in the mountains, nor for the friendships that these have generated, grateful though one must be for both. It is rather for the very special role that the series has played in integrating biomedical science in Europe. All of us know and respect science in the United States, and we have had good friends from the United States at these meetings. However, we need to feel proud of our scientific achievements on this side of the Atlantic, and this is just what these meetings have helped with so much. In particular I recall on one occasion hearing Vincent Eijsvogel describe here the very first evidence of recombination between class I and class II genes of the HLA complex (he was, I think, feeling his way towards this kind of nomenclature, although he used something more primitive at the time). Another memorable occasion was when Rolf Zinkernagel provided here the first rational explanation of the function of restriction, something that had previously become conspicuous for its absence!

My topic here is epitope linkage, the phenomenon whereby the immunological response to one epitope on a complex antigen becomes linked to the response to other epitopes. It is, I believe, one of the fundamental properties of the immune system. It explains how the immunolgical response manages to spread so effectively across and throughout a complex antigen. It also explains how the immune system has been able to develop the marvellous mechanism of hypermutation, which enables antibodies of the most exquisite specificity to be built. It underlies the complex pattern of interactions between lymphocytes, which in turn makes necessary the whole equipment of adhesion molecules and lymphokines that these cells bear, and which ultimately explains the architecture of lymphoid tissue. It provides a point of view from which to reinterpret immunological memory. It is the subject to which I have devoted my past 20 years, and which I have recently reviewed more formally and with references [1,2].

A general scheme of cell interactions in the immune system is presented in Fig. 1. On the left can be seen the well-established pathway of B cell activa-

K. Messmer M. Stein (Eds.)
Pathways in Applied Immunology

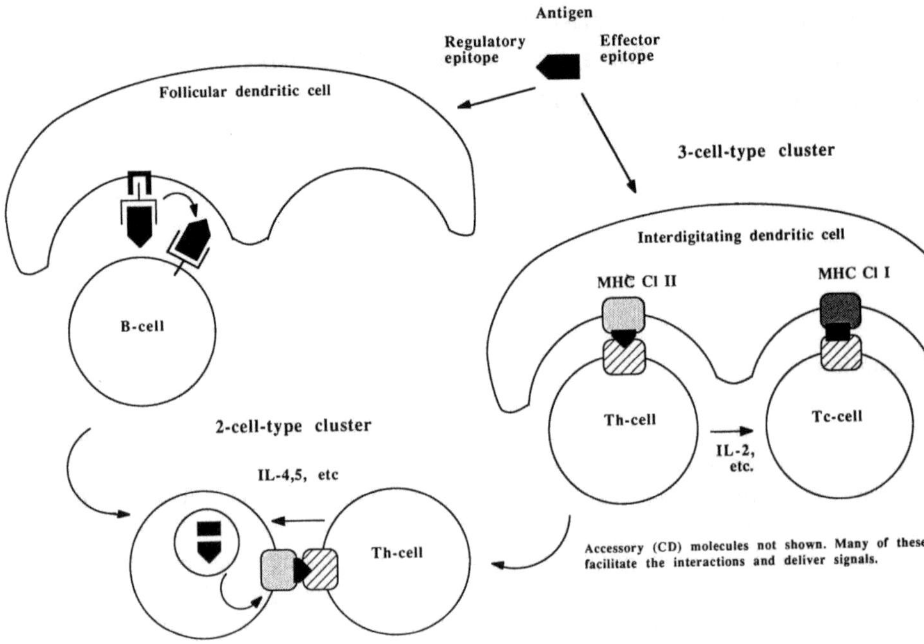

Fig. 1. Regulation of the immune response by cell clustering

tion, which starts with a B cell acquiring antigen from an antigen-antibody complex (also including complement components, not shown) held on the surface of a follicular dendritic cell. Right at the beginning of an immune response, before antibody begins to be secreted, B cells may acquire antigen directly, without drawing on the dendrite-bound stock. Next the B cell internalises the antigen within an endosome, where cleavage takes place mediated by cathepsin. Antigenic peptides released in this way bind to MHC class II molecules and are then translocated onto the B cell surface, where they are available to bind to helper T cells. The figure shows an activated T cell binding in this way to form a two-cell type of cluster or "conjugate". On the right can be seen an interdigitating dendritic cell (IDC) that processes antigen in much the same way as a B cell, ending up with peptides bound to both classes of MHC molecule. This enables the dendritic cell to form a three-cell type of cluster as shown, with cytotoxic-precursor as well as with helper T cells. Clusters of this type are less well documented than the T-B one, although there is some fairly direct evidence that they can form. The strongest argument is that something is needed to explain the various instances of epitope linkage in the cytotoxic response, and no better mechanism has been proposed.

Clusters are envisaged as delivering signals for the .activation of these effector cells, B cells and cytotoxic T cells, in the two ways illustrated in Fig. 2. Cytokines are synthesized and secreted much like any other secretory proteins. The final stage of their secretion by T cells appears to be markedly polarized,

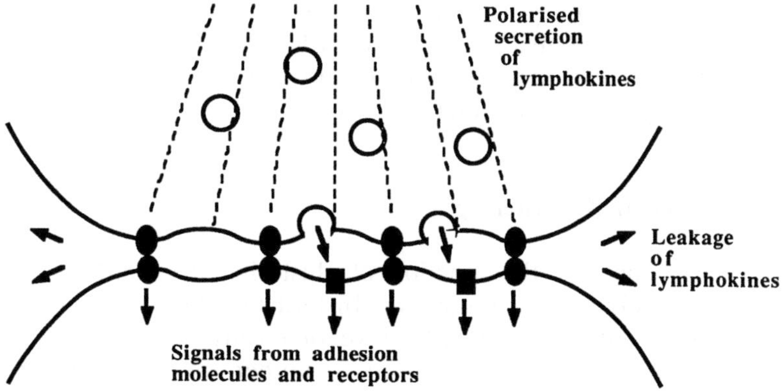

Fig. 2. Events at the cleft between cells interacting in a cluster

with exocytotic vesicles travelling down elements of the cytoskeleton that are oriented towards the area of contact with other cells in the cluster. Secretion is into a cleft, with the result that a high local concentration of the cytokine is achieved in the neighbourhood of its receptors. Some leakage undoubtedly occurs, as depicted in the figure, and this presumably has a physiological function in the case of the three-cell type of cluster, where adhesions form with the dendritic cell rather than with the cytotoxic-precursor cell that is the proper target of the cytokines. Signals are transduced not only by these receptors but also by the adhesion proteins that are responsible for forming and stabilizing the cleft. These adhesion proteins are of many different species, in addition to the MHC-T cell receptor (TcR) pair depicted in Fig. 1 above, and are often termed accessory molecules. Most and perhaps all such molecules transduce detectable signals, although for some species the signal may be no more than up-regulation of their own expression. Our information about most of the adhesion proteins is woefully incomplete; for instance, the natural ligand is often unknown, as in the case of CD45, and this is holding up progress. The sequence of events in clustering, and in the equally important process of declustering, is poorly understood. It is by no means clear that the MHC-TcR interaction is what starts the process off, and indeed the interaction between CD2 and ICAM-1 is at least as good a candidate. Declustering needs to take place smoothly, as after all the very great majority of clusters are unproductive, in the sense of the TcR not finding its appropriately antigenic peptide.

The sites at which these events take place are for the most part the T and B cell areas of lymph nodes and spleen. T-B clusters presumably form at the junction between T and B areas around the follicles in which the B cells have acquired their antigen. T-T-IDC clusters presumably form mainly within the cortex of lymph nodes, but some initial clustering seems to occur in peripheral tissue, as judged by the presence of clusters within afferent lymph. The origin of interdigitating dendritic cells (IDC) can be traced back to tissue dendritic cells, such as the Langerhans' cell of skin. Just how closely related these cells are to tissue macrophages is a matter of vigorous debate. My impression is that

IDCs are little more than macrophages that have taken up residence in particular environments, based on the difficulty of finding unique markers and on the absence for them of any counterpart to granulocyte/macrophage and macrophage colony-stimulating factors.

The Function of Clustering

One can consider the function of clustering from either a mechanistic or a broader evolutionary standpoint. In terms of the former, clustering serves the obvious function mentioned above of locally concentrating cytokines in the vicinity of their target receptors. Another obvious function is to organize communication between cells so as to enable a cytokine to reach an appropriate target cell. This lets a single cytokine, such as interleukin 2 (IL-2), be used on more than one kind of target, and to my mind this is the most likely reason why cytokines have been able to become so promiscuous. In addition, and much more speculatively, prolonged clustering might enable two or more cytokines to act in sequence, such as IL-4, IL-5 and IL-6 may do.

One other possible mechanistic function of clustering deserves special mention, in connection with the CD45 molecule that my colleagues are studying. This molecule has been discovered by others to have phosphatase activity in its cytoplasmic segment, and has been shown by them in man to be able to neutralize the signal transmitted by CD3, a phophokinase inducer. This finding we have extended to the mouse [2], and we have gone on to formulate the following hypothesis. Prior to clustering these two molecules, and perhaps other such pairs, presumably occur mixed together at random on the cell membrane, and this mixture would establish an equilibrium level of phosphorylation. Clustering would cause CD3 to aggregate at the cleft; this would locally perturb the equilibrium, and that would constitute a signal either on its own or have a modulating effect on other signals.

Concerning the broader background to these cooperative mechanisms, two reasons for their evolution have been suggested. One is that they represent a device for implementing epitope linkage. In Fig. 1 the antigen is depicted as bearing two epitopes, one regulatory and the other effector, and the cell interactions work so as to ensure that an effector response develops only when these two structures are physically linked. In effect, then, these mechanisms coordinate the response so as to make it spread across and throughout an entire antigenic structure. The structure can be far more than just a single protein, and although much of the experimental support has come from model systems using alloantigenic cells or artificial conjugates, it is reasonable to suppose that an entire virus or even an entire protozoan parasite would constitute a set of linked epitopes.

A second reason for this evolution is likely to have been that it permits hypermutation to have evolved. B cells, as is now well established, have evolved a unique mechanism that generates mutation in an expressed IgV region at an extraordinarily high rate. This rate may be as high as one mutation per gene per cell division, perhaps ten thousand-fold higher than normal. The generator appears to operate only in memory B cells, and therefore perhaps only in the environment of a secondary follicle. It appears to be highly advantageous, as it enables antibodies of high affinity to form ("maturation of the immune response", as the process has been called). Indeed the whole apparatus of follicular dendritic cells, with their Fc and C receptors, can be regarded as a device for storing conformationally intact antigen in order to drive this process of maturation. A mechanism of this sort is hardly compatible with the maintenance of self-tolerance by clonal deletion, as the gaps produced by deletion would rapidly become filled in by mutation. The immune system appears to have solved this problem by dividing in two, with one portion taking responsibility for self-tolerance, and the other then evolving hypermutation. Numerous observations support the view that T cells do indeed perform the first of these tasks: self-tolerance of F liver protein and C5 (intermediate concentration proteins) has been found in T but not B cells; T cells do not hypermutate.

Placed against this background, the leakage of cytokines out of the cleft as discussed above and depicted in Fig. 2 becomes a serious matter. It jeopardizes epitope linkage and self-tolerance and poses a threat of redundant effector activation. In order to minimize this threat the immune system appears to have evolved a number of mechanisms, of which perhaps the most important is the spacing well apart from one another of three-cell type clusters in the lymph node cortex. Indeed that has been my principal reason for identifying the IDC as the central cell in such clusters. Other mechanisms important for this purpose no doubt include rapid flow of lymph through lymphoid tissue, appropriate adjustment of lymphokine concentrations and stimulation thresholds, and so on. One interesting idea is that inhibitors may be important, both soluble and membrane bound. This might be the reason for the otherwise rather mysterious release of IL-2 receptors into solution from the lymphocyte surface. It might also be the reason why some 95% of IL-2 receptors are α-chain homodimers, unable to transduce a signal into the cell (a problem with this interpretation is that these homodimers have a lower affinity than the αβ receptors which do signal).

Antigen Storage and Lymphoid Architecture

Returning to Fig. 1, the type of antigen storage that is needed in the two halves is very different. On the left, what B cells need is conformationally intact antigen, because antibodies recognize the outer surface of antigens, and the shape of that surface depends largely (but of course not entirely) on the conformation of proteins. In addition, B cells need antigen to be available over a long time period, in order that it can drive their maturation via hypermutation. The retention of antigen in the form of antigen-antibody complexes on the surface of follicular dendritic cells (FDCs) can be regarded as a device for fulfilling these two requirements; indeed it is hard to imagine any other way of

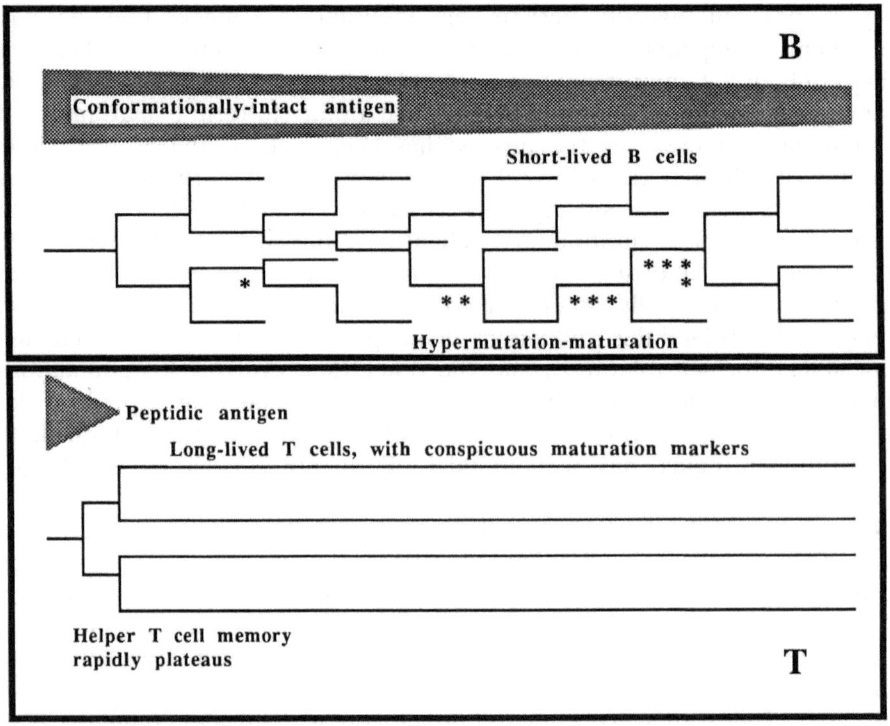

Fig. 3. Immunological memory has different components in the two lymphocyte compartments

doing so. On the right, T cells need something different. They recognize peptide fragments presented by MHC molecules, fundamentally because this ensures that they become activated only when in contact with the cells that are their targets. Because they do not hypermutate, there is no need for antigen to be available to them for prolonged periods. These differences in antigen storage and presentation are almost certainly the fundamental reason for using two cells as different as FDCs and IDCs to fulfil these requirements. And this in turn adequately explains why lymphoid tissue is partitioned into B and T cell areas.

Everything that we know about antigen storage fits this picture. Antigen can be detected for long periods, for months at least after administration, but only in conformationally intact form on FDCs. It can be visualized in this location by immunofluorescence, by autoradiography, or by functional tests using adoptive transfers or feedback inhibition. No doubt antigen continues to stimulate T cells via B cells and IDCs, but neither of these cells seems to be able to retain antigen for long.

Immunological Memory

As seen from this point of view, immunological memory has two components, one in B and the other in T cells. This division is illustrated in Fig. 3. Precisely because of the hypermutation-maturation process, we would not expect B cells themselves to be long lived, but rather the antigens on which they browse. This indeed seems to be the case, although the technology for measuring B cell survival is still fairly controversial. For T cells, on the other hand, we would expect a rapid response followed by a long plateau. That also seems to be the case; for instance, I have measured T cell memory of minor histocompatibility antigens in the mouse, and found just this (unpublished data). T cell memory is obviously useful, just as responsive T cells are themselves useful in the response. Accordingly we would expect to find long-lived memory T cells, and this too seems to be the case: witness Court-Brown's classic discovery of 8-year-old T cells after radiation treatment of ankylosing spondylitis.

A further point is worth making in this connection. Because of their rapid turnover we would not expect to be able to identify memory B cells as a particularly distinct population. For T cells the position is different, and we would expect to be able to identify a distinct population of hyperreactive memory cells. This indeed seems to be the case: the replacement of the CD45RA marker by CD45R0 is a conspicuous feature of such memory T cells in man, rat and mouse, i.e. wherever it has been looked for. Other markers have also been found on memory T cells, and this has become a fertile area of research.

References

1. Mitchison NA (1989) T cells in transplantation immunity. Immunol Lett 21:15
2. Fisher AG, Goff LK, Lightstone L, Marvel J, Mitchison NA, Poirier G, Stauss H, Zamoyska R (1989) Problems in the physiology of class I and class II MHC molecules, and of CD45. In: Proceedings Immunological Recognition Symposium, Vol 54. Cold Spring Harbor Laboratory, New York. pp 667–674

What We Have Learned from Highly Immunized Patients*

J.J. van Rood

I would like to begin by thanking Walter Brendel for initiating the Kitzbühel-Axams meetings. This talk is dedicated in his honour and in appreciation of what he has done. He has been in many ways a leader; he professionalised experimental surgery with the "big bang" machine as one of the many highlights, introduced us to xenotransplantation, but above all initiated these wonderful round table meetings on immunology. The recipe was as simple as it was brilliant: take some beautiful snow-covered mountains, mix them together, invite a lot of friends and ask them to talk about science. Most of these people have been involved in transplantation, but not all of them. We have heard talks about skin grafts in monkeys, about liver transplantation in pigs, and many, many other topics.

We have followed his initiative by organizing every year a similar immunology meeting on a boat. The boat has the big advantage over the mountains that one cannot get off. So there is always much discussion. It is now quite famous. The concept, however, is due to Walter's original idea; thus also on this level Walter has set an international example.

I would like to report here what we have learned in the past 2 years from highly immunized patients waiting for a renal transplant. Over 15% of the patients on the waiting list have leukocyte antibodies reactive with the lymphocytes of 85% or more of the donors. The main risk factors for the development of high sensitization are pregnancy, blood transfusion, and graft rejection. Highly immunized patients have antibodies against almost all foreign HLA antigens, and it is difficult to locate suitable transplants because cross-matches with almost all donors are positive. When these patients are treated like other patients with respect to donor selection and urgency, they accumulate on renal transplant waiting lists. Therefore, several strategies have been developed to increase the chance of finding a cross-match negative donor for them. One possibility is to select HLA-A and HLA-B identical or compatible donors, but the chance of finding such a donor is often slight because of the enormous polymorphism of the HLA system.

* This research was supported by the Netherlands Organization for Scientific Research (NWO), the I.A. Cohen Institute for Radiopathology and Radiation Protection (IRS), the Eurotransplant Foundation, and the E. Jung Foundation.

K. Messmer M. Stein (Eds.)
Pathways in Applied Immunology

Other schemes involve the distribution of sera from these patients to many tissue-typing laboratories. Each potential donor is tested against these sera, and by trial and error donor-recipient combinations with negative cross-matches are identified [1,2]. These strategies have contributed to a shorter waiting time for highly immunized patients.

We have developed a special strategy for managing highly immunized patients that does not require distribution of patient sera and can be operated from a laboratory attached to a central organ-sharing department. Our protocol is based on data obtained within Eurotransplant showing that, in the case of a negative cross-match with both current and historical sera HLA-A and HLA-B mismatches are less important for graft survival in highly sensitized patients, whereas matching fo HLA-DR has a significant influence, especially in retransplantation [3,4].

The aim of the strategy is exact definition of those HLA-A and HLA-B incompatibilities that do not result in a positive cross-match. The sera of highly immunized patients are screened against a panel of lymphocyte donors selected on the basis of having only one HLA-A or HLA-B mismatch with a specific patient. Using this approach, HLA-A or HLA-B antigens towards which antibodies did not develop can be detected in the patient. Kidney donor selection is made by entering data on the HLA-A, HLA-B, and HLA-DR antigens of all potential donors into a central computer, which then selects cross-match negative, highly sensitized recipients on the basis of their own HLA-A, HLA-B, and HLA-DR antigens in combination with the acceptable HLA-A and HLA-B mismatches [5]. In this way the donor is always matched for HLA-DR (either identical or compatible) but may have several HLA-A and HLA-B mismatches.

In the past 2 years, the sera from 73 highly sensitized patients were tested for the *absence* of specific alloantibodies against HLA-A and HLA-B antigens using such patient-specific panels. All 73 patients had antibody reactivity against more than 85% of the panel donors in a complement-dependent lymphocytotoxicity test. This antibody reactivity resulted from multispecific alloantibodies to HLA class I antigens; autoantibodies were excluded. In patients with less than 100% panel reactivity, a first indication concerning these acceptable HLA-A and HLA-B mismatches can be deduced from the HLA antigens of the negative panel donors in the screening.

We found, for 59 out of the first 73 patients tested, HLA class I antigens that, although different from those present in the recipient, did not lead to a positive cross-match. This approach has led to a significant increase in size of the potential kidney donor pool for these patients. We could identify for 34 of the 59 patients an HLA-DR compatible, cross-match negative kidney that carried one or more of the "acceptable" class I mismatches. These transplants did very well, with 29 grafts (85%) surviving for 1 year after transplantation (F.H.J. Claas et al., unpublished observations, 1988).

We realized later that the acceptable mismatches often included those HLA antigens of the patient's mother that the patient had not inherited (noninherited maternal antigen: NIMA). In 15 of the 26 patients of whom the

Table 1. Comparison of the NIMA haplotype to the acceptable mismatches of highly sensitized patients

Patient	PR[a]	Noninherited maternal haplotype	Acceptable mismatches include:
MI	100	A11 B35	A11 B35
JO	97	A1 B8[b]	A1
EN	95	A30 B35	A30 B35
VE	100	B7[c]	B7
KO	96	A3 B35	A3 B35
ST	100	A2[b] B27	B27
ZA	90	A1 B8[b]	A1
BO	98	A26 B40	A26 B40
BR	100	A3 B40[b]	A3
CA	98	A28 B44	A28 B44
NO	89	A2 B40[b]	B40
MO	98	B35[c]	B35
VM	98	A32 B27	A32 B27
KA	100	A2 B18[b]	B18
VG	100	B40[c]	B40

HLA class I typing was done by a standard NIH microcytotoxicity assay with 120 highly selected alloantisera. Sera were tested for panel reactive antibodies with a complement-dependent cytotoxicity assay. In brief, heat-inactivated serum (1 µl) was added to Ficoll-hypaque isolated mononuclear cells (4000) and incubated for 30 min at room temperature. Rabbit serum (5 µl) was added, the mixture was incubated for 60 min at room temperature, and 0.03% propidium iodide (5 µl) was added. The test was read on an automated inverted microscope. The panel reactivity (PR) was determined by testing the sera of patients against a panel of at least 50 lymphocyte donors selected so that all HLA-A and -B antigens were represented. The acceptable mismatches, that is, the HLA-A or -B alloantigens toward which the patient had not formed antibodies, were determined by testing the patient's serum against a panel of lymphocyte donors, who were mismatched for only one HLA class I antigen [5]. In 11 patients the acceptable mismatches did not include the antigens present on the not inherited maternal haplotype [17].

[a] PR, percentage of a donor panel reactive with the patient's serum.

[b] An unacceptable mismatch.

[c] Noninherited maternal HLA-A antigen is identical to paternal HLA-A antigen in the patient or mother is homozygous on HLA-A locus.

mother could be HLA typed, the acceptable class I mismatches included NIMAs (Table 1). Six of these patients had only one of the two NIMAs (either HLA-A or -B) classified as an acceptable mismatch, suggesting that not all HLA antigens have the same ability to induce or maintain nonresponsiveness. These patients formed antibodies aganist the NIMAs HLA-A2, -B8, or B40, which might be due to the strong immunogenicity of these antigens (F.H.J. Claas et al., unpublished observations, 1988). The polymorphism of the HLA system argues against the possibility that our findings are due to chance.

The noninherited paternal haplotypes were analyzed as a control; only 2 of the 25 noninherited paternal HLA-A or -B antigens (NIPA) tested were acceptable mismatches, which emphasized the preferential nonresponsiveness to NIMA ($p < 0.005$ by Fisher's exact test). One explanation of our results is that during fetal life maternal cells entering the blood stream of the fetus might be responsible for the induction of nonresponsiveness [6].

The assumption that true tolerance is maintained by chimerism [7] could explain why some patients had no antibodies to either the HLA-A and -B antigens on the noninherited maternal haplotype. However, chimerism can not account for the cases in which the nonresponsiveness is to only one class I antigen (for example, A1) and not the other (B8; Table 1, patient JO). In these patients an idiotype-antiidiotype network may play a role in the maintenance of the nonresponsiveness [8].

To study whether the NIMA have a similar influence on the T cell repertoire a limiting dilution analysis was performed to quantitate the cytotoxic T cell precursor (CTLp) frequency of children against their maternal cells. As a control CTLp frequencies were determined against the paternal cells. Such analysis was performed in 37 healthy children belonging to ten families. The CTLp frequencies against the NIMA were on average significantly lower than those against the NIPA. Three different CTL response patterns were found: in 17 of the 37 children CTLp frequencies against the NIMA and NIPA did not differ significantly; two had a significantly higher CTL response to the NIMA than the NIPA, and 18 showed significantly lower CTLp frequencies against the maternal cells compared with those against the paternal cells. More importantly, ten children in this last group and two children in the first group had hardly any detectable CTLp against their mother, whereas this was found in only two cases for the paternal antigens. This difference is statistically significant. The reproducibility and stability of the assay has been described extensively elsewhere [9]; furthermore, three of the present combinations have been tested twice on different occasions with similar results, emphasizing again the reproducibility of the assay. The difference in CTLp frequencies is thus not due to chance or variability of the assay, and they are not influenced by the age or sex of the children.

In view of these findings one would expect that maternal organ transplants would fare better than paternal ones, but most studies (e.g., [10]) have not shown a different survival of maternal kidney grafts compared to paternal grafts. This and similar studies can be criticized for not taking into account that female grafts overall do worse than male grafts [11]. On the other hand, the finding that antibodies against maternal HLA antigens can be detected in newborn children [12], the superior survival of maternal skin grafts, and the lower frequency of cellular rejection in maternal kidney grafts compared with those from paternal donors do support our hypothesis [13–15]. The observation that offspring of women with insulin-dependent diabetes mellitus (IDDM) have a lower risk of developing IDDM than offspring of men with IDDM [16] may be caused by similar induction of neonatal tolerance to the autoantigens of the maternal beta cells in the T cell repertoire of their children.

In conclusion, the present data show that not only self MHC and non-MHC but also, among other environmental factors, noninherited maternal HLA and perhaps also non-HLA antigens can contribute to the negative selection of the T cell repertoire. The interaction between fetus and mother may result not only in immunization of the mother by the HLA antigens of the fetus but also in partial education of the T cell repertoire of the child by the HLA antigens of the mother. The preferential reduction in donor-specific mixed lymphocyte reactions as observed after blood transfusions mismatched for the noninherited maternal HLA antigens may be based on a similar feto-maternal interaction. The present findings could have obvious implications for organ transplantation as well as the study of HLA and disease associations.

Acknowledgements. This work represents the combined efforts of E.L. Lagaaij, Li Zhang, and many others of our group but above all F.H.J. Claas.

References

1. Bradley BA, Klouda PT, Ray TC et al. (1985) Negative crossmatch selection of kidneys for highly sensitized patients. Transplant Proc 17:2465–2466
2. Schafter AJ, Hasert K, Opelz G (1985) Collaborative transplant study crossmatch and antibody project. Transplant Proc 17:2469–2471
3. Hendriks GFJ, de Lange P, D'Amaro J et al. (1985) Eurotransplant experience with highly immunized patients. Scand J Urol Nephrol 92:81–86
4. Klouda PT, Ray TC, Kirkpatrick J et al. (1987) Graft survival in highly sensitized patients. Transplant Proc 19:3744–3745
5. Claas FHJ, Gijbels Y, van der Velden-de Munck JJ et al. (1988) A special strategy to increase the chance of finding crossmatch negative kidneys for highly sensitized patients. Transplant Proc 20:947–948
6. Owen RD, Wood HR, Foord AG (1954) Evidence for actively acquired tolerance to Rh antigens. Proc Natl Acad Sci USA 40:420–424
7. Streilein JW, Gruchella RS, Wood P, Strome P (1982) Active clonal deletion in neonatal H2 tolerance. Ann NY Acad Sci 392:276–290
8. Streilein JW, Gruchella RS (1981) Analysis of neonatally induced tolerance of H-2 allo antigens. I Adoptive transfer indicates that tolerance of class I and class II antigens is maintained by distinct mechanisms. Immunogenetics 12:161–173
9. Zhang L, Guang Li S, Vandekerckhove B, Termijtelen A, van Rood JJ, Claas, FHJ (1989) Analysis of cytotoxic T cell precursor directed against individual HLA-A and -B alloantigens. J Immunol Methods 121:39–45
10. Cicciarelli J (1989) Living donor kidney transplants. In: Terasaki PI (ed) Clinical transplants 1988. UCLA Tissue Typing Laboratory, Los Angeles, pp 293
11. Takiff H, Cook DJ, Himaya NS, Mickey MR, Terasaki PI (1988) Dominant effect of histocompatibility on ten-year kidney transplant survival. Transplantation 45:410–415
12. Chardonnens X, Jeannet M (1980) Immunobiology of pregnancy: evidence for a fetal immune response against the mother(?). Tissue Antigens 15:401–406
13. Ivanyi P, Démant P (1964) Prolonged survival of maternal skin grafts in newborn rabbits. Folia Biol 11:321–323
14. Kois WE, Campbell DA, Lorber MI, Sweeton JC, Dafoe DC (1984) Influence of breastfeeding on subsequent reactivity to a related renal allograft. J Surg Res 37:89–93
15. Kalia A, Dobbins JG, Brouhard BH, Travis L (1988) Sex of the parental donor and cellular rejection of renal allografts in children. Transplantation 46:70–73
16. Warram JH, Krolewski AS, Kahn CR (1988) Determinants of IDDM and perinatal mortality in children of diabetic mothers. Diabetes 37:1328–1334
17. Claas FHJ, Gijbels Y, van der Velden-de Munck J, van Rood JJ (1988) Induction of B cell unresponsiveness to noninherited maternal HLA antigens during fetal life. Science 241:1815–1817

Immunological Enhancement

J.R. Batchelor

I would like to start by thanking Walter Brendel for the pleasure that he has given to all of us. We think of you, Walter, as a renaissance man: a scientist, a musician, and a lover of all that is beautiful. One of your most outstanding qualities is a generosity of spirit. It has been a great pleasure to know you.

I have been interested in immunological enhancement for probably too long; but nevertheless, I think it is an important subject. Its importance is connected with a question raised by Jon van Rood (this volume pp 43–47) who pointed out that many badly matched kidney grafts function very well. Every transplant surgeon knows that very heavy doses of immunosuppressive drugs must be used in the early weeks after kidney allografting. Later the amount of drugs required falls quite remarkably. This accounts for the fact that many transplants do very well. If it were necessary to maintain immunosuppression at the high level of the immediately post-operative period indefinitely, the results of transplantation would be very poor because of the toxic side effects of the heavy immunosuppression. There clearly are some mechanisms active in successful kidney transplant patients which modulate the patients' immune responses against the graft, and thus permit low-dose immunosuppression to be sufficient.

I believe that the mechanisms of long-term graft survival in animal models of immunological enhancement are similar to those protecting long-term human kidney allografts. An understanding of these mechanisms should enable us to induce unresponsiveness before transplantation takes place. This should allow us to use very low dose immunosuppression. It is now clear that long-term unresponsiveness to organ allografts induced in experimental animals, e.g. by immunological enhancement or short-duration treatment with immunosuppressive drugs, is due in part to the activity of suppressor cells. There have been many demonstrations in the literature of transplantation immunology of the adoptive transfer of suppression of graft rejection. The studies of Brent and colleagues [1], Monaco's group [2], my own laboratory [3,4], and Morris's group [5] have shown that suppressive activity can be transferred by T cells, and that it is specific.

My colleagues and I have postulated that the T cells which mediate these specific suppressor effects do not belong to a minority sub-population of T cells but are part of the familiar $CD4^+$ or $CD8^+$ T cell populations. We have suggested that they, like alloreactive T cells and T cells specific for nominal

K. Messmer M. Stein (Eds.)
Pathways in Applied Immunology
© Springer-Verlag Berlin · Heidelberg 1991

antigens, bind to an MHC–peptide binary complex [6]. To explain the specificity of suppression, it is suggested that the peptide present in the binary complex is derived from amino acid sequences of the V regions of the T cell receptors (TCR) of the clones causing allograft rejection.

This concept of suppressor T cells is, in essence, one of an idiotypic network of T cells. The question therefore arises as to what evidence exists that is consistent with this hypothesis? Several laboratories can be cited. Firstly, the studies of Damle and Engleman [7] have identified autoreactive $CD8^+$, $CD28^-$ T cells in humans which specifically suppress proliferative responses of defined, autologous, alloreactive $CD4^+$ T cells. The specificity of the autoreactive $CD8^+$ cells is consistent with the activating complex, being autologous class T MHC molecule plus a peptide derived from an idiotypic (V region) sequence of the TCR of the target T cells.

In experimental models, my own group has found that rats with long-surviving kidney allografts harbour T cells in their spleens which proliferate when co-cultured with X-irradiated, syngeneic T cell blasts purified from mixed lymphocyte cultures in which the stimulator cells are derived from the kidney allograft donor strain [8]. Kimura and Wilson have also described T cells in F1 rats which are specific for idiotypic variations in the TCR of parental strain rats [9]. Evidence from the laboratory of Wekerle [10] demonstrates that T cells can be isolated from Lewis strain rats which recover from experimental allergic encephalomyelitis, and these recovery-phase T cell lines specifically suppress the functions of syngeneic anti-myelin basic protein T cells in vivo and in vitro. The suppressive T cells do not inhibit syngeneic T cells specific for other nominal antigens.

It is, of course, acknowledged that the concept of suppressor cell specificity put forward here is still largely speculative. However it does have a number of attractions. It explains the specificity of suppression; as far as can be presently judged, there is no strong evidence which is in conflict; lastly, it suggests that suppressor T cells are like other T cells expressing $\alpha\beta$ TCRs in that they bind to an MHC molecule–peptide binary complex.

References

1. Kilshaw PJ, Brent L, Pinto M (1975) Suppressor T cells in mice made unresponsive to skin allografts. Nature 255:489–491
2. Monaco AP, Wood ML (1981) Models of specific unresponsiveness in adult animals: potential clinical application. Transplant Proc 13:547–555
3. Batchelor JR, Phillips B, Grennan D (1984) Suppressor cells and their role in the survival of immunologically enhanced rat kidney allografts. Transplantation 37:43–46
4. Chui YL, Batchelor JR (1985) Mechanisms underlying continued survival of rat kidney allografts after a short period of chemical immunosuppression. Transplantation 40: 150–153
5. Barber WH, Hutchinsion I, Morris PJ (1985) Mechanisms of kidney allograft maintenance in rats treated with cyclosporine. Transplant Proc 17:1391–1393
6. Batchelor JR, Lombardi G, Lechler RI (1989) Speculations of the specificity of suppression. Immunol Today 10:37–40

7. Damle NK, Engleman EG (1989) Antigen-specific suppressor T lymphocytes in man. Clin Immunol Immunopathol 53:S17–S24
8. Lancaster F, Chui YL, Batchelor JR (1985) Antiidiotypic T cells suppress rejection of renal allografts in rats. Nature 315:336–337
9. Kimura H, Wilson DB (1984) Antiidiotypic cytotoxic T cells in rats with graft-versus-host disease. Nature 308:463–464
10. Sun D, Win Y, Chluba J, Epplen JT, Wekerle H (1988) Suppression of experimentally induced autoimmune encephalomyelitis by cytolytic T-T cell interactions. Nature 332: 843–845

Induction of Peripheral Tolerance to Class I MHC Alloantigens in Adult Mice*

H. Wagner, K. Pfeffer, and K. Heeg

A major goal of transplantation immunology is to define gentle methods able to convert immune reactivity to transplantation antigens into a state of immune unresponsiveness. It is well known that this conversion is successful in neonates: introduction of foreign antigens into a developing immune system prevents the immune system from responding further to these antigens. Induction of unresponsiveness in the mature peripheral T cell pool, however, meets difficulties: receptor occupancy by antigens induces primarily sensitization rather than tolerance.

Evaluation of pretransplant transfusion effects have indicated that intravenous confrontation with allogeneic cells may induce specific immunosuppression rather than immunization [1,2]. One way of looking at this phenomenon is based on the veto cell concept pioneered by Miller, Bevan, and associates [3,4]. Unlinked to the specificity of their own antigen receptor, veto cells paralyze in vitro the response of T cells reacting to (MHC) antigens displayed by the veto cells [5]. Interestingly, upon intravenous transfusion of allogeneic lymphocytes [6] or even of MHC transfected cells [7], recipient mice develop a state of specific unresponsiveness – as if the transfused lymphocytes were veto cells.

The latter conclusion led us to analyze the efficacy of intravenous allogeneic cell transfusion to induce unresponsiveness in adult mice. The main findings have been the following:

1. Injected mice develop a long-lasting chimerism that is paralleled by an unresponsiveness to skin allografts.
2. Injected mice display a state of split tolerance, i.e., peripheral but not intrathymic cytotoxic T cell precursors (CTL-p) bearing high-affinity tolerogen-reactive T cell receptors (TCR/s) are clonally silenced.
3. $L3T4^+$ recipient T cells are necessary for induction of tolerance.
4. Donor-derived lymphocytes isolated from chimeric mice display effective veto functions.

Since technical details have been published in the September 1990 issue of the *Journal of Experimental Medicine*, we can proceed immediately to the

* Supported by Sonderforschungsbereich 322.

K. Messmer M. Stein (Eds.)
Pathways in Applied Immunology

findings. It should be noted that we are discussing the induction of tolerance only to class I MHC incompatible allogeneic cells. Addition of MHC class II incompatibility disparities means a change in the rules as presented here.

When adult B6-recipient mice are intravenously transfused with about 5×10^7 allogeneic bm1 splenocytes, the in vivo and in vitro reactivity of recipient mice changes dramatically. The majority (>65%) of B6-recipient mice tolerate bm1 skin grafts for more than 4 weeks. In vitro the frequency of anti-bm1 reactive CTL-p in the recipient spleen drops from about 1/500 (control) to 1/10000 within 8 days; the major loss of CTL-p frequencies takes place within 24 h. Selective reduction of anti-bm1 reactive T cells is not confined to CTL-p but also includes interleukin 2 producing T cells. Essentially similar results were obtained using recombinant mice differing selectively at the H-2D locus.

Frequency reduction of CTL-p does not affect thymic T cells, and primarily those peripheral CTL-p bearing high affinity TCRs for donor cells become silenced, i.e., CTL-p which are resistant to blockade with anti-CD8 monoclonal antibodies (mAb). Surprisingly, recipient mice depleted for $CD4^+$ T cells (by pretreatment with high doses of anti-CD4 mAb) become resistant to tolerance induction. This aspect is not fully understood, and helpful comments on this would be appreciated. It should be noted that in B6-recipient mice treated with cyclosporin A tolerance could be induced as effectively as in control mice; tolerance induction thus appears to be cyclosporin A resistant.

Using FITC-labeled anti-$H2D^d$ mAb we analyzed the state of chimerism in the B10.A → B10.BR mouse combination. While in bone marrow or thymus no donor cells were found, significant numbers (3%–9%) of D^d positive B10.A donor cells were detected for up to 20 days in the spleen of recipient B10.BR mice.

Using cell-sorter technology, we next analyzed whether upon in vitro depletion of recipient splenocytes from chimeric donor cells CTL responsiveness would reappear. The fact that unresponsiveness remained indicated that it was not due to active donor cell mediated suppression. Thus the profound reduction of CTL-p frequencies in (tolerant) recipient $Lyt2^+$ splenocytes must be explained by other mechanisms.

Even though donor $Lyt2^+$ T lymphocytes were found to be tolerant, i.e., clonally depleted for anti-recipient (anti-D^k) reactive CTL-p, the very same cells efficiently suppressed the primary activation of CTL-p with reactivity to donor class I MHC antigens. In other words, they displayed the hallmarks of veto cells. This suppression was antigen specific, since the reactivity to third-party antigens was not affected. It is to be noted that donor type cells from control mice do not exhibit veto activity. We thus conclude that by "parking" class I MHC incompatible donor cells via the intravenous route in adult recipients, these cells are activated to a state defined by the ability to veto recipient $Lyt2^+$ T cells to respond to donor MHC alloantigens.

Why do donor and recipient T cells not reject each other, or at least cause symptoms resembling those of a graft-versus-host disease? We have shown here that peripheral tolerance is not due to active suppression but rather to clonal paralysis of antigen-reactive $Lyt2^+$ CTL-p and interleukin 2 producing

Lyt2$^+$ T cells. Since donor lymphocytes exhibit clonal anergy to recipient class I MHC antigens, yet express remarkably efficient veto functions in vitro, we believe that the reciprocal peripheral tolerance in adult chimeric mice is caused, and maintained, by veto activity induced both in donor and in recipient Lyt2$^+$ T cells.

Opposing thymic deletion of antigen-reactive T cells as the sole basis for negative selection of antigen-reactive T cells [8], we believe that the veto cell system described here may represent a paradigm for natural peripheral tolerance to self antigens not present in the thymus (mechanism of peripheral tolerance). On the other hand, pretransplant transfusion represents a gentle procedure to "open the window" of unresponsiveness in adult mice to induce chimerism. If that chimerism could be stabilized, an indefinite state of transplantation tolerance against foreign MHC antigens would be possible.

References

1. Van der Kwast TH, Binanchi AJJ, Bril H, Benner R (1981) Suppression of antigraft immunity by preimmunization. I. Kinetic aspects and specificity. Transplantation 31:79
2. Faustman D, Lacy P, Davie J, Hauptfeld V (1982) Prevention of allograft rejection by immunization with donor-blood depleted of Ia-bearing cells. Science 217:157
3. Miller RG (1980) An immunological suppressor cell inactivating cytotoxic T-lymphocyte precursor cells recognizing it. Nature 287:544
4. Fink PJ, Shimonkevitz RP, Bevan MJ (1988) Veto cells. Annu Rev Immunol 6:115
5. Miller RG (1986) The veto phenomenon and T cell regulation. Immunol Today 7:112
6. Rammensee HG, Bevan MJ (1987) Mutual tolerization of histoincompatible lymphocytes. Eur J Immunol 17:893
7. Madson JC, Superina RA, Wood KJ, Morris PJ (1988) Immunological unresponsiveness induced by recipient cells transfected with donor MHc genes. Nature 332:161
8. Kiesielow P, Blüthmann H, Staerz UD, Steinmetz M, von Boehmer H (1988) Tolerance in T-cell-receptor transgenic mice involves deletion of nonmature CD4$^+$8$^+$ thymocytes. Nature 333:742

The Use of Donor-Specific Bone Marrow to Induce Specific Allograft Unresponsiveness (Tolerance) in Adult Animals Transiently Immunosuppressed with Polyclonal Antilymphocyte Serum

A. P. Monaco

It is a great pleasure and honor for me to speak on the use of polyclonal antilymphocyte serum and donor antigen, specifically donor-specific bone marrow, to produce specific unresponsiveness (tolerance) to allografted tissue and organs. This topic is especially significant in view of Prof. Walter Brendel's early and pioneering interests in the uses of polyclonal antilymphocyte sera in tissue transplantation. This presentation is taken from a recent summary [1] by my colleagues and me in which we updated the use of polyclonal antilymphocyte serum and donor-specific bone marrow to induce unresponsiveness to tissue and organ allografts in several species.

Indeed ever since the demonstration by Medawar and colleagues [2] of actively acquired immunological tolerance, the idea of transplanting organs without the use of chronic immunosuppression has been the ultimate goal of clinical transplantation. Current results of clinical organ transplantation utilizing chronic, nonspecific immunosuppression are truly extraordinary; the success of renal and extrarenal organ transplantation is such that organ replacement is perhaps the most effective and efficient method of treatment for a number of morbid and mortal diseases affecting man. Yet the complications of nonspecific immunosuppression – failure to control rejection, sepsis, spontaneous neoplasm, metabolic complications, drug toxicity – continue to be major concerns in clinical translantation.

General Considerations Applicable to Clinical Tolerance Induction

Method of Tolerance Induction

Obviously, it would be extremely desirable to transplant organs without incurring complications. It is conceivable that as methods of nonspecific chronic immunosuppression improve, it might not be necessary to induce specific unresponsiveness. This would occur only if the methods of inducing clinical unresponsiveness did not have a favorable risk-benefit ratio compared to chronic nonspecific immunosuppression. Thus, use of total-body irradiation to induce tolerance might incur greater long-term risk than chronic immunosuppression. Under such circumstances, chronic immunosuppression would be preferable to attempts to induce unresponsiveness. Selection of the best method

K. Messmer M. Stein (Eds.)
Pathways in Applied Immunology

of induction of tolerance in clinical organ grafting is obviously an important problem.

Definition of the Clinically Tolerant State, Complete Tolerance, and Unresponsiveness

Even if a suitably safe method of tolerance induction is available, a most important problem is the definition of the tolerant state. In experimental transplantation, specific unresponsiveness (tolerance) is easily defined. Experimentally, specific unresponsiveness is the survival of an organ or tumor allograft in a recipient without an ongoing requirement for continued nonspecific immunosuppression while the capacity to reject third-party allografts is maintained along with normal immune responses to other (bacterial, viral, etc.) antigens. Such a definition implies that this survival is associated with normal graft function and histology. This rigorous experimental definition of specific unresponsiveness is extremely difficult to extend to clinical transplantation at the present time. To define a patient as completely and specifically unresponsive (tolerant) would require that the putatively tolerant patient be taken off all immunosuppressive drugs while being expected to maintain normal graft function and histology and essentially all other immune responses. To this definition of complete tolerance must be added the parameter of time. Many transplant groups have had experience with one or more transplant recipients who have taken themselves off immunosuppression, and who have maintained normal graft function (within the limits of routine graft function testing) for several to many years before deterioration of graft function from chronic rejection became apparent. If an excellent recipient is slowly tapered off all immunosuppression as part of a tolerance-inducing protocol (or if a noncompliant patient takes himself or herself off immunosuppression) but still maintains normal function and histology for several years before beginning rejection, was the patient tolerant during this period? Or is this merely an example of very attenuated chronic rejection? In the absence of appropriate immunological monitoring to detect allograft immune reaction before the graft is injured, and deterioration of graft function becomes apparent, complete tolerance can only be defined as normal graft function and histology off all immunosuppression. Thus, the concept of time becomes an important factor, i.e., even if tolerance is established, there is no guarantee that it will be permanent. An excellent patient who is maintained chronically on low doses of immunosuppression, and who does not reject over many years, with minimal risk from low-dose immunosuppression is probably better served than a patient who is assumed tolerant, but who eventually rejects the graft several years after being taken off all immunosuppression.

Partial Tolerance or Unresponsiveness in Clinical Organ Transplantation

In experimental transplantation systems involving multiple donor-recipient histocompatibility differences, induction of unresponsiveness to some (but not all) histocompatibility antigen differences produces prolonged or attenuated survival, i.e., partial tolerance. It is likely that initial protocols to induce unresponsiveness clinically might be only partially effective, so that allograft response(s) to some but not all histocompatibility differences between donor and recipient are eliminated, leading to a modified (and therefore prolonged) survival of the graft. Clinical demonstration of this situation may be obtained in several ways. The partially tolerized recipients might show better graft survival (and/or function) for the same dose of immunosuppression that achieves less effect in patients not given the tolerance protocol. Alternatively, the same or better survival could be achieved with less immunosuppression in such patients. At the same time such improvements in survival with less immunosuppression would be accomplished with better preservation of immune responses to bacterial, viral, and other antigens, which would be further reflected by reduced morbidity, mortality, hospitalized day readmissions, costs, etc. It is very likely that the initial clinical attempts to produce unresponsiveness will achieve partial or incomplete tolerance reflected by the improved patient outcomes as defined above. Also, the ethical considerations and the ill-defined parameters of time (duration) of unresponsiveness that require maintenance of some low-dose immunosuppressive regimen in initial tolerance-inducing protocols will definitely require evaluation and measurement of clinical tolerance in terms of parameters associated with a partial tolerant state.

Immunological and Functional Measurements of Unresponsiveness

Obviously, if there were an excellent in vivo or in vitro assay which correlated with recipient anti-donor alloreactivity, reduction and/or elimination of nonspecific immunosuppression could be obtained with more confidence that harm was not being done to the patient and graft. Also, during the time period when immunosuppression is reduced or eliminated (and graft function appears to be maintained), a good test for absence of alloreactivity would be reassuring that reduced or absent immunosuppression can be safely maintained. Unfortunately, such assays are as yet not available (the mixed lymphocyte reaction, MLR, not withstanding). Efforts to develop appropriate in vitro correlates of in vivo clinical alloreactivity should be emphasized as an integral part of any program to induce clinical tolerance. Furthermore, since good assays for alloreactivity are lacking, we should put additional emphasis on better tests of serial graft function and histology to monitor the graft condition during the putatively tolerant state.

Donor-Specific Antigen and the Induction of Tolerance (Unresponsiveness)

Various aspects of induction of tolerance in adult animals have been reviewed by the author previously [3]. Billingham et al. [2] induced tolerance (so-called actively acquired immunological tolerance) by infusion of donor antigen (living lympoid cells) into neonatal mice during their natural immunosuppressed state in the neonatal period. The principles to be used in tolerance induction in clinical transplantation were essentially defined by this unique experiment. Immunologically competent adult recipients would have to be transiently non-specifically immunosuppressed and then exposed to donor-specific antigen. Experimentally, the tolerizing antigen could be presented in the form of the graft itself, in a cell-free form or in the form of living-dissociated, replicating nongraft cells. In certain weak rodent histocompatibilty combinations, an allograft itself may induce tolerance with transient immunosuppression or even without immunosuppression. It is possible that some well-matched recipients of clinical allografts who have never rejected under chronic immunosuppression may indeed be tolerant. Unfortunately in the absence of an assay to define unresponsiveness, the only way in which this could be proven is to take the patient off immunosuppression, with the problem that late rejection may occur as defined above. Cell-free extracts of tissues have been used to induce various degrees of unresponsiveness in experimental animals. Although such extracts have worked to varying degrees in rodents, they have been significantly ineffective in large animals, particularly in whole-organ, vascularized allografts. In contrast, suspensions of replicating cells, especially lymphoid cells, have been extremely effective in inducing unresponsiveness to various types of allografts in various experimental animals including larger species using whole-organ, vascularized allografts [3].

Use of Donor-Specific Bone Marrow and Polyclonal Antilymphocyte Serum for Induction of Specific Unresponsiveness (Tolerance) to Experimental and Clinical Organ Allografts

The ALS–Bone Marrow Model in Mice

Monaco and colleagues (especially M. Wood, T. Maki, and J. Gozzo) first described the use of donor-specific bone marrow as the optimal cellular antigen in combination with short-term, nonspecific immunosuppression using polyclonal antilymphocyte serum (ALS) to induce unresponsiveness to experimental skin allografts (for reviews see [3–5]). A standard experimental protocol has been used in mice in which immunosuppression with ALS is given before and after test skin allografting followed by intravenous injection of donor-specific cells 1 week after transplantation. Using this standard protocol cell-free antigens as well as nonreplicating epidermal and hepatic cells were relatively ineffective. However, bone marrow cells were the most effective and efficient cellular lymphoid antigen in inducing specific unresponsiveness; some 20%–30% of marrow-treated animals maintained grafts over 100 days, while all ALS-treated mice (no marrow) rejected grafts in 30–40 days. Dose and timing of bone marrow infusion were important; the maximal effective dose of marrow was 25 $\times 10^6$–50 $\times 10^6$ cells/mouse, and maximal tolerance was achieved when the threshold effective dose was given about 1 week after grafting. However, at higher doses also spleen cells were effective. Recent experiments suggest that depending on the dose of ALS used, bone marrow may be given 1 or 2 days after grafting and still be effective.

Although the intravenous route was most effective, it was not obligatory; direct intrasplenic or intrahepatic injection of bone marrow into recipients produced unresponsiveness. Treatment of recipients with various bacterial adjuvants before or after bone marrow infusion could augment or obliterate the tolerance-inducing effect of bone marrow. These latter two observations suggested that the underlying mechanism of unresponsiveness was a positive immune response rather than one of clonal deletion. Finally, although the animals rendered tolerant by this method were not gross chimeras by standard criteria, experiments with chromosome markers suggested that donor bone marrow cells persisted in very low numbers for long periods in the spleens of unresponsive recipients. Although this observation was initially puzzling, subsequent observations have provided explanations for this (see below).

Mice rendered tolerant to donor skin grafts using ALS and bone marrow have been studied by in vitro assays to elucidate and identify the mechanism of tolerance. Lymphoid cells removed from such mice show normal responses to B cell and T cell mitogens. Cells from ALS-treated, marrow-infused mice fail to show lymphocyte-mediated direct cytotoxicity in vitro as long as the grafts are not rejected. Tolerant recipient lymphocytes fail to respond to donor cells in MLR. However, if the latter cells are reexposed to donor targets after MLR incubation, excellent cytotoxicity is achieved. Thus, cells from tolerant ALS-treated, marrow-infused mice do not proliferate in MLR but can generate a cell-mediated lymphocyte cytotoxicity on reexposure. One interpretation of these findings is that a suppressor-cell mechanism might be operative in producing and maintaining tolerance. Utilizing a standard suppressor cell transfer assay we subsequently showed that excellent tolerance could be transferred by lymphocyte cells from tolerized animals to ALS-treated syngeneic hosts. Administration of cytoxan, a known inhibitor of suppressor cells, prior to bone marrow infusion abrogated the tolerance usually induced by ALS and bone marrow. This provided further evidence that a suppressor cell mechanism was operative in this model.

We have utilized an extensive series of coculture assays in vitro to document what appears to be a multistep process in the evolution of suppressor cells in this model. In these studies, lymphocyte cells were removed from ALS-treated marrow-infused mice at various stages after treatment, and their ability to inhibit a recipient (C57BL/6) anti-donor (C3H/He) MLR in vitro was examined. Administration of ALS alone, without allografting or bone marrow infusion induced the transient production of nonspecific suppressor cells capable of suppressing a standard MLR in vitro. Addition of a test skin allograft, in this case a C3H/He graft on an ALS-treated B6AF recipient led to production of a donor-specific host-derived suppressor cell capable of suppressing only the specific C3H/He-B6AF MLR in vitro. Although this cell is detected for a longer time, it is detectable only during the time that the graft remains unrejected. If donor-specific C3H/He bone marrow is added, a third type of suppressor cell, present first in the spleen but later in lymph nodes, is subsequently detectable which is bone marrow derived (from the donor) and capable of suppression in recipient-donor MLR in vitro. Thus, a continuum of suppressor cell development – recipient-derived nonspecific, recipient-derived specific, and donor-derived specific – underlies the basic mechanism in this unresponsiveness system. This latter donor-derived cell which suppresses response against itself has by definition the function of a "veto" type cell.

Donor bone marrow has been fractionated by a number of standard methods (unit gravity, Percoll and Ficoll gradients, etc.), primarily through the work of J. Gozzo and S. De Fazio in our group, and an active fraction has been obtained which is equally as effective as whole bone marrow in producing unresponsiveness in ALS-treated mice. The active fraction (less than 10% of the whole marrow cell volume) is composed essentially of all medium-sized lymphocytes. Phenotypic marker analysis of this active fraction has shown that active marrow cells that prolong skin graft survival in ALS-treated mice appear to be Ia^-, Thy 1^-, largely complement receptor-negative, and Ig^-, but largely positive for Fc receptors. The active bone marrow cells may be natural suppressor or natural regulatory cells (veto cells?).

From the experimental and clinical point of view, uses of active bone marrow fractions are desirable since they avoid the use of large volumes of bone marrow, and there is some evidence experimentally that contaminating erythrocyte elements may exert a countereffect on tolerance induction. The active fraction can be frozen and preserved and then thawed without losing its effectiveness. The active fraction can be obtained from peripheral blood lymphocytes and from splenocytes although the best source in rodents has been bone marrow. Current experiments are now focusing on the possibility that this active bone marrow cell can be replicated and expanded in vitro for use in in vivo tolerance induction. Indeed, recent studies by T. Maki in our group suggest that the tolerance-inducing bone marrow cell is interleukin (IL-3) dependent. IL-3 dependent bone marrow clones have been produced with capacity to inhibit donor anti-recipient MLRs in vitro and prolong skin allografts in ALS-treated mice in a way similar to whole bone marrow.

Two recent experiments are of great importance. Since cyclosporin A (CyA) is so effective in clinical transplantation, it will be very difficult to try a tolerance-inducing protocol clinically which does not incorporate CyA initially and for a significant portion of time after transplantation. Addition of a brief course of CyA after bone marrow infusion in the standard ALS–bone marrow protocol in mice markedly enhances the number of animals rendered tolerant as well as the duration of tolerance [6]. This is presumably secondary to the enhancing effect of CyA on suppressor cell development. These studies emphasize that ALS–bone marrow protocols can easily and ethically be incorporated in clinical protocols currently using CyA. A recent study using artificial donor-specific urethane sponge grafts which are coated with donor-specific MHC antigens by a brief period of intraperitoneal residence in donor strain animals has shed additional insight into the mechanism of tolerance in this model. Such grafts can be removed at various times from animals made tolerant with ALS and bone marrow and their cell content analyzed. Such sponge allografts collect suppressor cells with donor specificity. This finding introduces the concept of a local form of tolerance and is in keeping with important observations made by the Thomas's in the rhesus monkey adaptation of the rodent ALS–bone marrow model [7].

The ALS–Bone Marrow Model in Dogs

Various degrees of unresponsiveness have been demonstrated in large animals with ALS and bone marrow using whole-organ, immediately vascularized renal allografts. Monaco and colleagues originally showed that ALS-treated dogs enjoyed prolonged survival of renal allografts following infusion of donor-specific bone marrow [8]. Subsequent studies showed that the active marrow fraction demonstrated in rodents could be isolated from canine marrow and used effectively to induce specific unresponsiveness to donor-specific canine renal allografts. Frozen, preserved active canine marrow fractions were also effective. Thus all the aspects of the ALS–bone marrow model in rodents have been reproduced in canines. Indeed in one experiment a third-party renal allograft was rapidly rejected while a well-tolerated donor-specific renal allograft continued perfectly in place after 1 year's survival. Recent studies have shown that addition of a short course of CyA after ALS and bone marrow treatment also increases the degree (number) and duration of tolerized animals in the canine renal allograft model as it does in rodents [9]. Median survival time for all canine allografts given ALS and bone marrow is superior to that in dogs given only ALS; however, only 30%–40% of animals show long-term tolerance, and these long-term survivors have varying degrees of chronic rejection, reflected in slightly elevated creatinine levels and abnormal biopsies. Thus, not all long-term renal allograft recipients are completely tolerant in this model. Most recent studies have shown that a longer (but finite) course of CyA after bone marrow produces a further increase in the number of dogs rendered tolerant (W.C. Hartner, unpublished observations).

The ALS–Bone Marrow Model in the Rhesus Monkey

The Thomas's [7,10,11] have adapted the rodent and animal ALS–bone marrow model to the rhesus monkey. Their elegant studies have shown the prolonged acceptance of MHC-mismatched kidney allografts in rhesus monkeys given posttransplant rabbit antithymocyte serum and donor-specific bone marrow. They have also shown that long-term surviving rhesus renal allografts after RATG and donor-specific bone marrow are infiltrated with nodular collections of lymphocytes. A significant portion of these lymphoid cells have donor specificities and suppressor cell phenotypes. Thus, a local suppressor cell effect (local tolerance) is also suggested as part of the unresponsive mechanism in the rhesus monkey. They have shown that along with the dramatic prolongation of median survival time following treatment with RATG and donor-specific bone marrow there is a dramatic reduction in incidence of anti-donor lymphocyte-mediated cytotoxicity (LMC) but a preservation of capacity to exert anti-donor antibody-dependent cellular cytotoxicity (ADCC) reactions. These investigators have also studied the effect of adjunctive immunosuppression in this model [10]. Azathioprine decreased the observed median survival time and increased anti-donor mixed lymphocyte culture (MLC) responses. In contrast low-dose CyA and prednisone increased the median survival time by 50% and prevented detectable LMC. However, CyA and prednisone appeared to increase the humoral alloimmune response, increasing the incidence of recipients with anti-donor ADCC. Long-term survival in this group was limited by chronic rejection and CyA toxicity. Their findings point out that success in tolerance induction is not an all-or-none phenomenon. An additional concept also suggested by their studies is that improved success in this system may require adjunctive immunosuppression that is directed to the humoral alloimmune response [10].

Donor-Specific Bone Marrow Infusion and Polyclonal Antilymphocyte Treatment in Human Renal Allografts

Initial Trials

The ALS–bone marrow model has been effective in producing various degrees of unresponsiveness in all species tested (both large and small) and in both nonimmediately and immediately vascularized allografts. Monaco et al. [12] performed an initial trial of bone marrow infusion in living-related, MLC-positive combinations in the pre-CyA era. This was an uncontrolled trial, and although the results were encouraging (only two acute rejections in the first 100 patient months), they were also inconclusive, and evidence for chronic rejection in long-term survivors was present. This later study emphasized that lack of a good in vitro assay of immune unresponsiveness. Long-term survivors after ALS–bone marrow treatment had specific suppression of their MLC versus the specific donor while third-party reactivity was restored. Even in the absence of donor-specific MLC reactivity, evidence of chronic rejection was present (creatinine, 2.0–2.5 range) in spite of very prolonged survival and the absence of acute rejections. These patients seemed to exhibit only partial tolerance as described above. Currently, there is an on-going, controlled trial in primary cadaver transplantation, and the initial results are encouraging [13], but clearly more, detailed trials are necessary.

Considerations for Future Trials

Donor bone marrow is the ideal donor cellular antigen for widespread trials in clinical organ allografting. Multiorgan donors can have their organs harvested, and donor ribs could subsequently be removed for use as the bone marrow source. In the original trials [12,13] aspiration of iliac crests was done. In the case of multiorgan donor cadavers prolonged harvesting from iliac crests can lead to hemodynamic instability and could cause the loss of some potentially harvestable organs. Use of donor ribs (after all harvests are completed) is simple and does not interfere with the technical aspects of harvesting. Bone marrow derived from ribs could be fractionated to its active fraction which could be frozen and preserved for infusion 1–2 weeks later depending on the timing of bone marrow infusion. Isolated, preserved donor-specific marrow could be distributed to recipients from a central processing laboratory. Several other potential sources of tolerance-inducing donor-specific antigen may also be useful. Rodent experiments show that the spleen can provide the tolerogenic cell. Current studies in our laboratory suggest that fractionated peripheral blood lymphocytes contain the tolerogenic cell in small numbers. A very interesting possibility is that the tolerogenic cell could be isolated from donor marrow and expanded in culture for use 1 or 2 weeks after harvest. If the ALS–bone marrow protocol is utilized for all recipients of a multiorgan donation, it may be necessary to expand the available marrow to achieve sufficient donor-specific marrow for all recipients.

The lack of availability and/or reproducibility of current polyclonal ALS preparations may be a problem. Experiments are now being done in our laboratory to determine whether current monoclonal anti-T cell antibodies (OKT3, for example) can be used to replace the polyclonal sera tested experimentally. As noted above, a simple standard protocol of grafting, post-transplantation ALS treatment, along with prednisone and azathioprine followed by bone marrow infusion and addition of CyA treatment would be easy to incorporate into current immunosuppressive protocols. Many units use polyclonal ALS, prednisone, and azathioprine for 7–14 days and then cycle in CyA. Donor bone marrow infusion could be a simple addition to these protocols for both renal and extrarenal organs.

Although the ALS–bone marrow protocol would be easy to incorporate into current multidrug immunosuppressive programs, interpretation of a salutory effect attributable to bone marrow infusion might be difficult. All of the above experiments suggest that pure tolerance as an all-or-none phenomenon will not be produced. Rather, some type of partial tolerance will be achieved. Thus, marrow-infused patients in a multidrug program might do better in patient and graft survival than similarly treated multidrug patients *not* given bone marrow. Also, differences could be measured by the ease with which various drugs in the maintenance programs of bone marrow and control groups can be subtracted, decreased, or totally eliminated, hopefully demonstrating that marrow infused patients could be maintained with less administration of drugs and with less frequent rejection reactions. The likely drugs to be subtracted initially are steroids. If steroids can be successfully subtracted, another drug could then be removed. It is important to emphasize that complete removal of all immunosuppressive drug therapy, although a commendable and worthy goal, is not likely to be achieved rapidly. Furthermore, as noted above, as drugs are subtracted, one cannot necessarily rely on persistent normal function. Successfully prolonged grafts must be serially monitored carefully for function to detect any deterioration at the earliest time so that immunosuppression could be reinstituted. Indeed periodic biopsy surveillance of organ allografts subjected to tolerance-inducing programs will have to be added to the monitoring program until a noninvasive in vitro or in vivo assay for clinical tolerance is forthcoming.

References

1. Monaco AP, Wood ML, Maki T, Gozzo JJ (1989) Future strategies in immunosuppression: Problems and potential for the induction of specific unresponsiveness to organ allografts in clinical transplantation. Transplant Proc 21:3939
2. Billingham RE, Brent L, Medawar PB (1953) Actively acquired tolerance to foreign cells. Nature 172:603
3. Monaco AP (1988) Biological immunosuppression: Polyclonal antilymphocyte sera, monoclonal antibody and donor specific antigen. In: GJ Cerilli (ed), Organ transplantation and replacement. Lippincott, Philadelphia, p 83
4. Monaco AP, Wood ML, Maki T, Hartner W, De Fazio S, Gozzo JJ (1988) Cyclosporine and unresponsiveness to allografts induced by polyclonal ALS and donor specific bone marrow. Transplant Proc 20:36
5. Monaco AP, Wood ML, Maki T, Gozzo JJ, De Fazio S (1988) The use of donor specific antigen for the induction of immunological unresponsiveness to experimental and clinical allografts. Transplant Proc 20:122
6. Wood ML, Gottschalk R, Monaco AP (1988) The effect of cyclosporine on the induction of unresponsiveness in ALS-treated, marrow injected mice. Transplantation 46:449
7. Thomas JM, Carver M, Cunningham P, Park K, Gonder J, Thomas F (1987) Promotion of incompatible allograft acceptance in rhesus monkeys given post-transplant ATG and donor bone marrow. I. In vivo parameters and immunohistologic evidence suggesting microchimerism. Transplantation 43:332
8. Caridis T, Liegeois A, Barrett I, Monaco AP (1973) Enhanced survival of canine renal allografts in ALS treated dogs given bone marrow. Transplant Proc 5:671
9. Hartner WC, De Fazio S, Maki T, Markees TG, Monaco AP, Gozzo JJ (1986) Prolongation of renal allograft survival in ALS treated dogs by postoperative injection of density gradient fractionated donor bone marrow. Transplantation 42:593
10. Thomas JM, Carver M, Foil MB et al. (1983) Long term incompatible kidney survival in outbred higher primates without chronic immunosuppression. Ann Surg 198:370
11. Thomas JM, Carver M, Cunningham P, Sash C, Park K, Thomas F (1989) Promotion of incompatible allograft acceptance in rhesus monkeys given post-transplant ATG and donor bone marrow. II. Effect of adjuvant immunosuppressive drugs. Transplantation 47:209
12. Monaco AP, Wood ML, Maki T, Madras PN, Sahyoun AI, Simpson MA (1985) Attempt to induce unresponsiveness to human renal allografts with ALG and donor specific bone marrow. Transplant Proc 17:1312
13. Barber WH, Diethelm AG, Laskow DA, Deierhoi MA, Julian BA, Curtis JJ (1989) Use of cryopreserved donor bone marrow in cadaver kidney allograft recipients. Transplantation 47:66

Mixed Bone Marrow Reconstitution Across MHC Barriers

D.H. Sachs, C.H. Chester, and M. Sykes

Introduction

It has been known for some time that reconstitution of lethally irradiated mice with allogeneic bone marrow leads to chimerism and tolerance across major histocompatibility (MHC) barriers [1,2]. However, if mature T cells are not removed from the allogeneic bone marrow, this procedure leads also to graft-versus-host disease (GVHD), which severely limits the applicability of this approach. If all mature T cells are first removed from the allogeneic bone marrow inoculum, a graft can still be achieved in mice, but in large animals, including humans [3] and miniature swine [4,5], such depletion produces major problems of failure of engraftment. The reason for this difference between rodents and large animals probably involves the much higher tolerance which mice exhibit to the toxic effects of total body irradiation (TBI) as compared to the larger species. Mice are routinely prepared for bone marrow transplantation by administration of 10 Gy TBI at 1.0 Gy/min, and the only target tissue damage which appears to be fatal is that of the lymphohematopoietic system. TBI administration at similar dose and dose rate to large animals, on the other hand, produces unacceptable levels of toxicity to other target organs, such as the gut and the lungs [6,7]. In order to develop approaches to achieving alloengraftment across MHC barriers, therefore, it is important to examine preclinical models in which T cells are not depleted from the allogeneic inoculum. We review here some of our own work with such models, including attempts to diminish the GVHD induced by allogeneic T cells without causing failure of engraftment.

Materials and Methods

The animals were male mice – C57BL/10SnJ(B10), B10.D2/nSn(B10.D2), B10.BR/SgSn(B10.BR), and A/J – obtained from Jackson Laboratories (Bar Harbor, Maine) or from Frederick Cancer Research Facility (Frederick, Maryland). Animals were maintained in a specific-pathogen-free facility.

Chimeras were prepared as previously described [8]. Briefly, recipient B10 mice, aged 12–16 weeks, were lethally irradiated (10.25 Gy) and reconstituted on the same day with bone marrow obtained from the femora and tibiae of B10

K. Messmer M. Stein (Eds.)
Pathways in Applied Immunology

and B10.D2 donors aged 6–12 weeks. T-cell depletion (TCD) was performed as previously described using rabbit anti-mouse brain (RAMB) serum and guinea pig complement (Gibco, Grand Island, NY) [9] or using anti-CD4 (GK1.5) or anti-CD8 (2.43) monoclonal antibodies (MAb) plus rabbit complement [10]. Animals were reconstituted with 5×10^6 syngeneic TCD B10 plus 15×10^6 allogeneic non-TCD B10.D2 or A/J bone marrow cells.

FITC-conjugated MAb 5F1 [11] was used for detecting cells of B10 origin in chimeras. Biotinylated MAb 34-2-12 [12] with specificity for D^d was used for detection of cells of B10.D2 or A/J origin. FITC-conjugated anti-K^k MAb 36-7-5 [13] was also used for detection of cells of A/J origin. FITC-conjugated and biotinylated Leu4 MAb were purchased from Becton Dickinson and were used as negative controls. Fluorescence microfluorometry analyses were performed in one- and two-color modes as described [8] using a modified Becton-Dickinson FACS 2. Initial peripheral blood lymphocyte phenotyping was performed 6–10 weeks after bone marrow transplantation as described [8].

Results

Completely Allogeneic Chimeras. Administration of bone marrow inocula from which only the host bone marrow T cells were depleted led to reconstitution of recipients as completely allogeneic chimeras [8,14]. Such animals appeared to be relatively protected from GVHD by the coadministration of TCD syngeneic marrow in the reconstituting inoculum [14]. However, the ability of TCD syngeneic marrow to diminish GVHD mortality was observed only during the early posttransplant period, when control animals died of acute GVHD, and not during the later period, when mortality due to chronic GVHD was observed. Thus, early deaths appeared to be prevented, but eventually all animals succumbed to GVHD [14].

Effect of IL-2. During subsequent studies of the in vivo anti-GVHD effect of cells cultured from TCD syngeneic marrow in the presence of IL-2 [23], we observed that control animals treated only with interleukin 2 (IL-2) also appeared to be protected from GVHD. This finding has provided the basis for a new therapeutic protocol for prevention of lethal GVHD [15]. Administration of high doses of IL-2 to lethally irradiated mice during the first 5 days following irradiation and reconstitution with allogeneic bone marrow plus spleen cells led to marked improvement in survival. The optimal anti-GVHD effect was observed when both TCD syngeneic bone marrow and IL-2 were administered, as shown in Fig. 1. Completely allogeneic reconstitution resulted in all survivors, regardless of whether or not TCD syngeneic bone marrow cells were coadministered.

Immunocompetence of Completely Allogeneic Chimeras. We have attempted to evaluate the immune function of long-term survivors after administration of allogeneic spleen plus bone marrow cells along with syngeneic TCD marrow

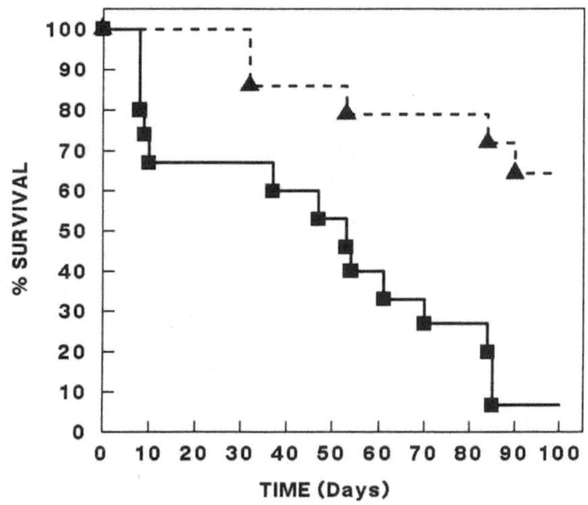

Fig. 1. Effect of IL-2 and TCD syngeneic marrow on GVHD mortality from A/J lymphocytes. Lethally irradiated B10 mice received 8 × 10^6 A/J spleen cells plus 1.5 × 10^7 A/J bone marrow cells with (▲---▲) or without (■—■) 5 × 10^6 TCD B10 bone marrow cells and IL-2, 50 × 10^3 U peritoneally twice daily for 5 days

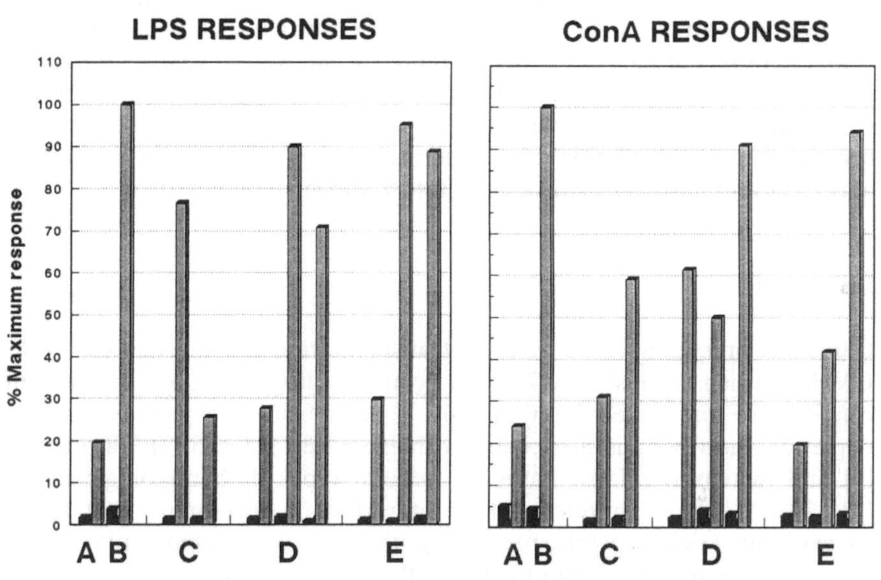

Fig. 2. Responses of completely allogeneic chimeras to LPS and ConA. Mouse spleen cells from individual animals of the following groups were cultured with (*shaded bars*) or without (*solid bars*) the indicated mitogen: *A*, Normal B10; *B*, normal A/J; *C*, B10 → B10 control chimeras; *D*, A/J bone marrow plus spleen cells → B10, plus IL-2; *E*, A/J bone marrow plus spleen cells plus B10(-T) → B10, plus IL-2

and protective doses of IL-2. The majority of these studies have been conducted on animals which survived longer than the 100-day follow-up period in experiments involving the A/J → B10 model, which represents a strong allogeneic difference, and leads to severe GVHD. Control mice not treated with IL-2 generally died early from acute GVHD in these experiments, so that such controls were not available for evaluation of long-term immune function. One of the major causes of chronic GVHD (cGVHD) related morbidity in mice is immunodeficiency [16]. The manifestations of cGVHD-related immunodeficiency in mice include reduced proliferative responses to the T cell mitogen concanavalin A (ConA) [17,18] and to the B cell mitogen lipopolysaccharide (LPS) [18,19]. Results of experiments examining the responses of these completely allogeneic chimeras to LPS and to ConA are presented in Fig. 2. These date indicate that most long-term IL-2 treated survivors demonstrate ConA and LPS responses which are comparable to those of normal animals.

Effect of Selective Depletion of the Syngeneic Component. Previous data indicated that depletion of mature T cells from both the syngeneic and allogeneic components of the mixed inoculum led to stable, mixed lymphohematopoietic chimerism [9,21], while depletion of only the syngeneic inoculum led to complete allogeneic chimerism [14]. To determine which subpopulations of the syngeneic component must be depleted to permit complete allogeneic chimerism to develop, groups of animals were prepared in which the allogeneic marrow component of the inoculum was untreated while the syngeneic component was depleted selectively with anti-CD4, with anti-CD8, or with both reagents. As indicated in Table 1, depletion of $CD4^+$ cells only led to syngeneic reconstitution, but depletion with either anti-CD8 alone or the combination of anti-CD4 and anti-CD8 led to completely allogeneic chimerism.

Discussion

Clinicians practicing bone marrow transplantation are faced with a dilemma. GVHD, which is caused by T lymphocytes in allogeneic bone marrow inocula, can best be prevented by removing T cells from the marrow [3,21]. Such depletion, however, is associated with increased rates of engraftment failure [3] and increased relapse rates for several types of hematologic malignancy [21, 22]. We have therefore been interested in developing new approaches to avoiding GVHD while preserving the graft-versus-leukemia (GVL) and alloengraftment-promoting effects of allogeneic T cells.

Previous studies from our laboratory demonstrated that coadministration of TCD syngeneic marrow is associated with a delay in acute GVHD mortality compared to that observed in recipients of similar allogeneic spleen cell/bone marrow cell inocula without TCD syngeneic marrow [14]. During attempts to culture cells from the syngeneic marrow which might be responsible for this protective effect, we adoptively transferred IL-2-cultured cell populations to mice which would otherwise have been expected to develop GVHD [23].

Table 1. Selective T cell depletion of the syngeneic component of a mixed bone marrow inoculum[a]

Group	Treatment of bone marrow	Animal number	B10 (%)	B10.D2 (%)	Graft rejection (days)	
					B10.D2	B10.BR
1	Anti-CD8	939	<5	>95	>100	27
		942	<5	>95	NT	NT
		947	<5	>95	NT	NT
		2001	<5	>95	>100	26
		2010	5.1	NT	NT	NT
2	Anti-CD4	941	89.3	<5	10	10
		945	86.6	<5	7	12
		946	93.1	<5	10	NT
		948	>95	<5	10	10
		949	>95	<5	10	10
		2007	91.7	<5	7	10
3	Anti-CD4 plus Anti-CD8	937[b]	<5	>95	>43	NT
		940[b]	<5	84.6	>72	>72
		943	9.7	>95	>100	NT
		950	<5	>95	NT	NT
		2002	<5	>95	>100	23
		2003	<5	>95	>100	NT
		2006	<5	>95	NT	NT
		2051	<5	>95	>100	23

[a] Strain combination: B10(−T) + B10.D2(+T) → B10.
[b] Animal died with skin graft intact.

We administered IL-2 to these mice in vivo in the hope of augmenting any protective activity. Surprisingly, the control animals treated with IL-2 but not receiving injections of IL-2-cultured cells were also protected from GVHD mortality. This observation led to the results reviewed here, which indicate the IL-2, when administered to mice in sufficient doses early after bone marrow transplantation, has a potent ability to protect against both acute and chronic GVHD mortality, and that the addition of IL-2 treatment led to much greater protection than was provided by TCD syngeneic marrow alone [15].

Because one of the major causes of morbidity and mortality in patients suffering from cGVHD is infection, which occurs commonly because of the immunodeficiency caused by cGVHD [16], we have also examined the responsiveness of completely allogeneic chimeras to mitogens. Our data indicate that, at least as far as responses to LPS and con A are concerned, long-term survivors protected from acute GVHD by IL-2 and TCD syngeneic marrow show essentially normal reactivity, suggesting the absence of cGVHD. Studies of additional parameters of immunity in these animals are in progress.

Studies of selective TCD of the syngeneic component of mixed inocula were carried out to determine whether all mature T cells need be eliminated to achieve completely allogeneic chimeras. Surprisingly, we found that depletion of only the $CD8^+$ T cells was sufficient for this purpose. These data suggest either that $CD4^+$ T cells do not contribute to allogeneic resistance in this fully MHC-mismatched combination, or that there are too few $CD4^+$ T cells in mouse bone marrow to be effective in preventing alloengraftment. If $CD4^+$ T cells can remain in the syngeneic marrow component without preventing allogeneic engraftment, and if these cells are capable of further mitigating GVHD, then these findings may be relevant to future protocols aimed at maximizing anti-GVHD effects in this model.

References

1. Beverley PCL, Brent L, Brooks C, Medawar PB, Simpson E (1973) In vitro reactivity of lymphoid cells from tolerant mice. Transplant Proc 5:679
2. Rayfield LS, Brent L (1983) Tolerance, immunocompetence, and secondary disease in fully allogeneic radiation chimeras. Transplantation 36:183
3. Martin PJ, Hansen JA, Torok-Storb B, Durnam D, Przepiorka D, O'Quigley J, Sanders J, Sullivan KM, Witherspoon RP, Deeg HJ, Appelbaum FR, Stewart P, Weiden P, Doney K, Buckner CD, Clift R, Storb R, Thomas ED (1988) Graft failure in patients receiving T cell-depleted HLA-identical allogeneic marrow transplants. Bone Marrow Transplant 3:445
4. Popitz-Bergez FA, Sakamoto K, Pennington LR, Pescovitz MD, McDonough MA, MacVittie TJ, Gress RE, Sachs DH (1988) Bone marrow transplantation in miniature swine. II. Effect of selective genetic differences on marrow engraftment and recipient survival. Transplantation 45:27
5. Suzuki T, Sundt TM, Kortz EO, Mixon A, Eckhaus MA, Gress RE, Spitzer TR, Sachs DH (1989) Bone marrow transplantation across an MHC barrier in miniature swine. Transplant Proc 21:3076
6. Deeg HJ, Storb R, Thomas ED (1984) Bone marrow transplantation: a review of delayed complications. Br J Haematol 57:185
7. Freirich EJ, Gehan EA, Rall DP, Schmidt LH, Skipper HE (1966) Quantitative comparison of toxicity of anti-cancer agents in mouse, rat, hamster, dog, monkey, and man. Cancer Chemother Rep 50:219
8. Sykes M, Sheard M, Sachs DH (1988) Effects of T cell depletion in radiation bone marrow chimeras. I. Evidence for a donor cell population which increases allogeneic chimerism but which lacks the potential to produce GVHD. J Immunol 141:2282
9. Ildstad ST, Wren SM, Bluestone JA, Barbieri SA, Sachs DH (1985) Characterization of mixed allogeneic chimeras. Immunocompetence, in vitro reactivty, and genetic specificity of tolerance. J Exp Med 162:231
10. Sykes M, Chester CH, Sundt TM, Romick ML, Hoyles KA, Sachs DH (1989) Effects of T cell depletion in radiation bone marrow chimeras III. Characterization of allogeneic bone marrow cell populations that increase allogeneic chimerism independenty of graft-vs-host disease in mixed marrow recipients. J Immunol 143:3503
11. Sherman LA, Randolph CP (1981) Monoclonal anti-H-2Kb antibodies detect serological differences between H-2Kb mutants. Immunogenetics 12:183
12. Goodman JW, Burch KT, Basford NB (1972) Graft-vs-host activity of thymocytes: relationship to the role of thymocytes in hemopoiesis. Blood 39:851
13. Goodman JW, Basford NL, Shinpock SG, Chambers ZE (1978) An amplifier cell in hemopoiesis. Exp Hematol 6:151
14. Ildstad ST, Wren SM, Bluestone JA, Barbieri SA, Stephany D, Sachs DH (1986) Effect of selective T cell depletion of host and/or donor bone marrow on lymphopoietic

repopulation, tolerance, and graft-vs-host disease in mixed allogeneic chimeras (B10 + B10.D2 → B10). J Immunol 136:28
15. Sykes M, Romick ML, Hoyles KA, Sachs DH (1990) In vivo administration of interleukin 2 plus T cell-depleted syngeneic marrow prevents graft-versus-host disease mortality and permits alloengraftment. J Exp Med 171:645
16. Sullivan KM, Witherspoon RP, Storb R, Anasetti C, Appelbaum FR, Bigelow C, Clark J, Doney K, Hill R, Loughran T, Matthews DD, Nims J, Petersen F, Sanders J, Schuning F, Shields A, Strom S, Thomas ED (1989) Chronic graft-versus-host disease: recent advances in diagnosis and treatment. In: Gale RP, Champlin R (eds) Bone marrow transplantation: current controversies. Liss, New York, p. 511
17. Seddick M, Seemayer TA, Lapp WS (1984) The graft-versus-host reaction and immune function. I. T helper cell immunodeficiency associated with graft-versus-host-induced thymic epithelial damage. Transplantation 37:281
18. Holda JH, Maier T, Claman HN (1985) Murine graft-versus-host disease across minor barriers: immunosuppressive aspects of natural suppressor cells. Immunol Rev 88:87
19. Iwasaki T, Fujiwara H, Shearer GM, Iwasaki T (1986) Loss of proliferative capacity and T cell immune development potential by bone marrow from mice undergoing a graft-vs-host reaction. J Immunol 137:3100
20. Ildstad ST, Sachs DH (1984) Reconstitution with syngeneic plus allogeneic or xenogeneic bone marrow leads to specific acceptance of allografts or xenografts. Nature 307(5947): 168
21. Poynton CH (1988) T cell depletion in bone marrow transplantation. Bone Marrow Transplant 3:265
22. Butturini A, Gale RP (1988) T cell depletion in bone marrow transplantation for leukemia: current results and future directions. Bone Marrow Transplant 3:265
23. Sykes M, Hoyles KA, Romick ML, Sachs DH (1990) In vitro and in vivo analysis of bone marrow-derived $CD3^+$, $CD4^-$, $CD8^-$, $NK1.1^+$ cell lines. Cell Immunol 129:478

Liver Transplantation

R. Calne

Introduction and Historical Background

The ultimate goal in organ transplantation is to produce immunological tolerance so that the patient is specifically rendered incapable of reaction against the organ in question but is not otherwise immunologically paralysed. To do this without the use of poisonous immunosuppresive drugs would certainly be preferable, but for the moment it is likely that the only way in which we can achieve this goal in the short term is by a preliminary conditioning period using the agents that we already have and those that are likely soon to become available. Despite chemical immunosuppressive drugs having been available for 30 years it is still not clear exactly how they should be used, although it is apparent that the dose must be tailored individually for each patient in question. We know that there are natural inhibitory control mechanisms of the immune system, and it is likely that in our approach to immunosuppression we are not able to take account of natural controls and probably destroy suppressor mechanisms in the course of our therapy. Probably the requirements of immunosuppression vary at different stages in relation to the timing of the transplant and the reaction to it.

At the time of the first Brendel Meeting held in Kitzbühel in 1969 there were many exciting developments in transplantation immunology, and our meetings were held with a Parliament-style seating arrangement so that discussion involved eye contact with other participants. This seemed to me to lead to active and spirited discourse. In the intervening years we have tended to specialise so that we learn more and more about less and less. Increasing complexity arises from what initially had seemed simple.

Nevertheless, in order to modify the immune system it may be necessary to have a much clearer mechanistic concept of how the human species reacts to an allograft, and perhaps one day to a xenograft. The liver allograft is an especially interesting model since rejection can be more easily controlled than with other organs in all species. In the rat a liver allograft engages the immune system in an early conflict which results in a relative de-cloning of the lymphocytes which react against it, then the recipient becomes operationally tolerant towards the donors major histocompatibility complex (MHC). This seems also to occur in the pig. In the rat and the pig it is not always necessary to give immunosuppression.

K. Messmer M. Stein (Eds.)
Pathways in Applied Immunology

New chemical agents such as FK506 and biological materials, in particular "humanised" monoclonal antibodies, which can be given for much longer periods, may enable us to produce tolerance in the clinic. For the present, however, we must still use relatively toxic compounds. It has been found that giving small doses of each compound combined, the immunosuppressive effect can be additive with virtual elimination of most of the serious individual side effects, and this triple or quadruple therapy has now become the most popular approach to clinical immunosuppression.

Our experience in liver transplantation goes back to 1968. Shortly after beginning in Cambridge we formed a joint programme with Roger Williams' liver unit at King's College Hospital in London. We followed Starzl et al. [5] who had done the first clinical liver transplant in 1963. The early results were dreadful, both in Denver and in Cambridge, but then a few successes occurred, and it became clear that some patients could do well. Whenever it is possible to show in surgery that an operation can mean excellent therapy, that operation tends to come within the routine armamentarium of surgical procedure, but it may take a few years. It has taken a long time for liver transplantation, but there has recently been a tremendous increase in activity. There are more than 70 centres in North America and in Europe, with nearly 2000 livers being grafted each year. We have now done nearly 900 liver transplants since 1968, and in the course of these 23 years many changes have been made in patient selection technique and management. Some patients have done very well, surviving more than 16 years. The longest survivor in the world is a patient of Starzl's who has survived more than 20 years since transplantation. Although liver transplantation does not confer immortality, for the majority of patients now it is an excellent therapy, especially when the alternative is inexorable worsening of the fatal liver disease unless a transplant is performed. Very young patients are being transplanted although the risks in those aged under 1 year and those aged over 60 years are much higher than in older children and younger adults.

Indications

Any fatal disease confined to the liver is potentially an indication for liver transplantation. In addition there are a group of diseases which lead to death due to enzyme defects of the patient's otherwise normally functioning liver. The cases fall into the following four broad categories.

1. Primary Malignancy of Which Primary hepatoma is the Main Cause. Unfortunately, the results are bad because of recurrent tumours from micrometastases undetectable at the time of transplantation. Approximately 60% of patients die within 2 years after transplantation from recurrent tumours. Nevertheless, there have been some long-term survivors, including our longest case of 16 years.

2. *Cirrhotic Diseases*. The most common indications have been primary biliary cirrhosis and chronic active hepatitis. Alcoholic cirrhosis is an indication if it appears that the patient is unlikely to resume drinking. Assessment of alcoholics is difficult; some have given up drinking because they are too ill, others really wish to stop altogether. Starzl feels that a liver transplant operation is the ultimate deterrent to a resumption of alcoholism, and of the cases reported only a minority have returned to drink.

3. *In born Errors of Metabolism*. Unlike the previously considered cases, following liver transplantation in in-born errors of metabolism cannot recur, nor can biliary atresia. In Wilson's disease and alpha$_1$-antitrypsin deficiency the indication for surgery is liver failure. In other conditions such as oxylosis the liver is macroscopically normal, but is unable to produce an essential enzyme. In oxylosis patients die of renal failure, and they may need a combined kidney and liver transplant; we have performed 10 such operations, and 7 are doing well with their combined kidney and liver grafts. It would be atrractive to use hepatocytes or an accessory extra liver to provide the missing enzyme, but unfortunately we know from the work of Starzl and his colleagues in the 1960s that an extra liver in the presence of an otherwise normally functioning native liver atrophies in about 2 weeks and shrivels up to be a useless small piece of tissue. The only way to get an extra graft to survive is to compromise the patient's own liver by interrupting the portal supply or damaging the liver parenchyma. Thus an orthotopic liver graft removing the patient's own liver is generally felt to be the best surgical option.

4. *Acute Fulminating Liver Failure and Sub-acute Hepatic Necrosis*. This may occur as primary diseases or follow a liver transplant that has acute rejection or compromised blood supply. These patients withstand surgery well but they are difficult to manage since by the time the diagnosis of liver necrosis has been made the patient has only a short time to survive and it may be very difficult in that time to obtain a donor liver. Whilst waiting the patient may develop permanent brain damage which is difficult to assess.

Technique and Management

Details of the technique have been published and I will make only a few comments in relation to the operation.

Timing. Acute liver failure leaves little option, and the co-operation of colleagues in other centres is necessary to obtain donor organs before the patient dies on account of irreversible brain damage. In some cases the liver must be removed, for instance, in gross bacterial sepsis or trauma even without a donor liver being available, a porto-caval shunt being performed and the patient maintained in the intensive care ward until hopefully a new liver can be found within 1–2 days. The first successful case was performed in Hannover. The logistics of liver transplantation have improved following introduction of a new preservation solution in which glutathionate and raffinose are the most important impermeate ions. This solution was first used by my colleague Neville Jamieson when he was working with Professor Belzer in Wisconsin [4]; in Cambridge now, a new simplified development of the solution appears to be producing excellent results with preservation in the clinic up to 24 h. Not only is it more pleasant for the surgical team to be able to schedule the operation for a morning start, but it is an obvious advantage for the recipient not to have an exhausted team following the operation.

Care. Care of the patient during surgery is extremely important. We are fortunate in the support of the Anaesthetic Department. It is far better to anticipate complications, especially bleeding, and treat them before a severe and possibly irreversible deterioration has occurred. In adult patients we use venous shunting in only about 10% of cases; usually patients with cardiac or renal insufficiency are deemed to be especially at risk. The technique that we use is a modification of that described in Pittsburgh, Pennsylvania, in which the inferior vena cava is decompressed through the saphenous vein and the portal system through the inferior mesenteric vein (Fig. 1). The blood is then returned

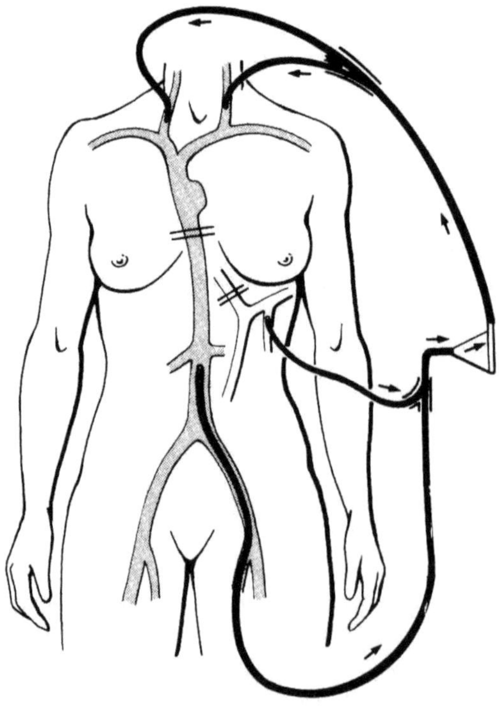

Fig. 1. Venovenous bypass. A canula is inserted into an IVC through the saphenofemoral junction and into the inferior mesenteric vein. A Biomedicus pump without heparin is used to re-infuse the blood through the neck canulas inserted by the anaesthetist

Fig. 2. Patient with combined heart, lung and liver transplant now 3 years since the operation

to the superior caval system through the anaesthetist's neck lines by a Bio-medicus pump and no systemic heparin.

Blood Requirements. Blood requirements in a liver transplant can vary from 0 to more than 100 units; the average loss in Cambridge is around 10 units for adults. In children there is a severe shortage of donors; because of the size requirements it is obviously impossible to put a large liver into a confined space. It may be necessary to use a lobe from a donor who would otherwise be too large, a technique pioneered in Hannover and Paris. Now most centres reduce the liver in some cases. The most difficult part of the surgery is usually removing the diseased liver, particularly if there has been previous upper abdominal surgery and shunting procedures. There are now many examples of multiple organ transplants with the liver as one of the organs. Patients who have received such grafts tend to have done well. Results are thought to have been better than transplanting the organs alone. The only long-term survivor of a functioning intestinal transplant was grafted with a liver in London, Ontario [3]. The first patient to have a combined heart, lung and liver transplant in Cambridge is well at 3 years without evidence of rejection of any of the organs (Fig. 2) [6].

Aftercare and Follow-up. The infra-structure of support in liver transplantation is absolutely vital. The surgery is only one part of the care of the patient. The intensive care facility needs to be adequate in bed space and quality of nursing staff. We do not begin to administer cyclosporin before there is good evidence of good renal function. We begin our immunosuppression with azathioprine and steroids, sometimes with added anti-lymphocyte globulin. The relationship of tissue typing and liver transplantation is not clear, and reports have been conflicting. The two largest series, namely that in Pittsburgh and our own, have reported negative correlation between outcome and tissue matching [1,2]. Since these observations have been retrospective, we now have in progress a prospective trial in Europe with centres in Hannover, Paris and Cambridge co-operating with J.J. van Rood's group in Leiden in a study of the relationship of tissue match to outcome. A negative correlation might be explicable in terms of viral diseases, especially those which affect the liver and auto-immune diseases, but any comment on this would be premature and speculative. Certainly, recurrence of hepatitis B virus replication is a serious worry in patients who have the virus prior to surgery, and we do not yet have a good way of controlling this. The best results have been obtained by giving large repetitive doses of hyper-immune gamma globulin.

Conclusion

I conclude with a word of tribute and homage to my old friend Walter Brendel, a scientist of outstanding vision and humanity, whose pupils have founded their own departments of surgery and experimental surgery that have restored the ancient prestige of German surgery to the highest levels. This memorial is entirely appropriate to a very great man.

References

1. Donaldson PT, Alexander GJ, O'Grady J, Neuberger J, Portmann B, Thick M, Davies H, Calne RY, Williams R (1987) Evidence for an immune response to HLA class I antigens in the vanishing bile duct syndrome after liver transplantation. Lancet i:945–948
2. Gordon RD, Fung JJ, Iwatsuki S, Duquesnoy RJ, Starzl TE (1988) Immunological factors influencing liver graft survival. Gastroenterol Clin North Am 17(1):53–59
3. Grant D, Wall W, Mimeault R, Zhong R, Ghent C, Garcia B, Stiller C, Duff J (1990) Successful small-bowel/liver transplantation. Lancet i:181–184
4. Jamieson NV, Sundbert R, Lindell S, Laravuso RB, Kalayoglu M, Southard JH, Belzer FO (1988) Successful 24–30 hours liver preservation: a preliminary report. Transplant Proc 20:945–947
5. Starzl TE, Marchioro TL, Von Kaulla KN et al. (1963) Homotransplantation of the liver in humans. Surg Gynecol Obstet 117:659–676
6. Wallwork John, Williams Roger, Calne RY (1988) Transplantation of liver, heart and lungs for primary biliary cirrhosis and primary pulmonary hypertension. Lancet i:182–184

Pancreatic Islet Transplantation

P.J. Morris

In choosing the topic of pancreatic islet transplantation, I return to a subject of research activity in our group that dates from 1974, when I went to Oxford. It was soon thereafter that I first attended a symposium at Axams upon Walter Brendel's invitation.

For those who are not particularly involved in looking after diabetic patients the major problem of insulin-dependent diabetes is the secondary complications of diabetes: the retinopathy, the nephropathy, the peripheral vascular disease, and the coronary artery disease. As a result of all these complications, apart from the immediate problems that they produce in these patients, the insulin-dependent diabetic has a life expectancy some 15 years less than those without the condition. This then leads to the question of whether these secondary complications can be prevented by better control of the glucose abnormalities that are a feature of insulin-dependent diabetes. Indirect evidence suggests that the better the control of glucose levels through a 24 h period by insulin administration, the less likely one is to get these complications, however the evidence for this is not absolutely watertight. And indeed we do not know why, for example, among 100 juvenile insulin-dependent diabetics only 50 ever develop any of the secondary complications. This is obviously genetically determined, but we are not yet able to determine which patients will develop the secondary complications, and this is relevant to the rationale of transplantation of the endocrine pancreas.

The rationale for pancreatic transplantation is that a successful transplant of the endocrine pancreas would completely correct the glucose and other abnormalities, primarily the former, and thereby prevent the development of these secondary microvascular complications of diabetes. Again, there is experimental evidence in animals and some not very convincing evidence in humans that this is indeed the case. The two approaches to transplantation of the endocrine pancreas are firstly to transplant the whole pancreas or a segment of the pancreas as a vascularised organ, which has been practised for some years and now very successfully in a number of centres, such as Munich, and secondly to transplant isolated pancreatic islets or fetal pancreas. The reason that we have remained firmly wedded to the possibility of transplanting pancreatic islets is that it is unlikely to be possible to transplant a vascularised pancreas at an early stage of the disease, which probably will have to be done if one is to prevent the development of these vascular complications. At the

K. Messmer M. Stein (Eds.)
Pathways in Applied Immunology

moment vascularised pancreatic transplantation must be confined to patients with end-stage secondary vascular complications and usually requiring a kidney transplant for diabetic nephropathy, as well. There are exceptions to this, but by and large one can never expect a vascularised pancreatic graft to be transplanted, for example, in a 15-year-old who has had diabetes for 2 years with the prospect of life-long immunosuppression thereafter. Now, if pure islets could be transplanted, it would be a relatively safe procedure, and in addition there is the theoretical possibility of perhaps modulating the immune response to these grafts by pretreating the islets themselves.

The technique used for most experimental work in the rodent was that described by Lacy and Ballinger in 1970. Rat pancreases were digested with collagenase to obtain the islets, but four to six pancreases are needed to obtain enough islets. The islets were separated over a density gradient and then injected into the portal vein. This technique was particularly successful and produced very pure islets in the rat, and when injected into the liver, they survived very nicely and immediately corrected the diabetes in a streptozotocin-treated rat (which destroys the β-cells). After transplantation of syngeneic islet grafts into the portal vein, a diabetic rat within 24 h has a normal blood glucose. Diabetic rats produce an enormous volume of urine (in excess of the body weight over 24 h), but this returns to normal after islet transplantation, and they start to regain weight again.

When we started this programme we went on to develop other methods of isolation and other sites for implantation of islets and showed, for example, that they can be put into the spleen very satisfactorily or, indeed, under the kidney capsule. The advantage of putting them under the spleen or under the kidney capsule is that the spleen or kidney can be removed with the islets implanted, and these animals can be made diabetic again, confirming that the islet transplants were responsible for the correction of the diabetes. That is quite an important point because it has been well recognized that even doses of streptozotocin which uniformly produce diabetes do in fact, as time goes by, allow the survival of some β-cells and in some instances spontaneous correction of diabetes.

These were the sorts of models that we developed about 10–12 years ago to look at some of the immunological aspects of islet transplantation in the rat, and I will wish to focus now on only one or two areas of current interest. One is the rejection of allogeneic islets. A number of factors influence the rejection of islets; this I shall not discuss, but in general they behave as if they were very immunogenic, very much like a skin allograft. For example in the DA to Lewis rat strain combination with no treatment at all, allogeneic islets are rejected as judged by recurrence of diabetes (blood glucose over 200 mg%) within 4 days. The same applies to implantation in the spleen although under the kidney capsule islets are not rejected for about 6–7 days. Furthermore, a wide variety of immunosuppressive protocols – antilymphocyte globulin, passive enhancement, cyclosporine in very large doses, and cyclophosphamide – produced only a modest prolongation of survival. This represents a summary of many experiments performed in our laboratories about 10 years ago.

It proved very difficult to produce prolonged survival of isolated pancreatic islets in the rat model; one recalls that in the same strain combination kidneys and hearts can be prolonged indefinitely by all these immunosuppressive protocols. This is one of the problems that we will obviously be faced with in clinical transplantation. It is interesting, however, that islets transplanted as part of a vascularised pancreas do not appear to be particularly immunogenic, because in fact some patients reject the kidney and not reject the pancreas. Most of the rejection reaction of the vascularised pancreas seems to be directed more towards the exocrine than the endocrine tissue. Indeed, some years ago we developed a model in which islets removed from the donor animal were parked in an intermediate syngeneic host under the kidney capsule for 24 or 48 h, before transplanting the kidney with its parked islets as a vascularised organ. Now with immunosuppression the islets and the kidney were not rejected. In other words, when presented to the host as a vascularised allograft, rejection could be suppressed relatively easily.

One interesting experiment I would like to describe was designed to look more closely at the rejection reaction, because we have often wondered why islets are rejected so readily in these rodent models – which is the universal experience, I should point out. There has also been substantial interest in the diabetic field in the specific damage caused by lymphokines, in particular interleukin 1 (IL-1), to pancreatic β-cells. In particular, Nerup from Copenhagen has shown in vitro that human islets and rodent islets are very susceptible to damage by IL-1 and indeed has produced some evidence that there are in fact IL-1 receptors on pancreatic β-cells. He has suggested that IL-1 may play a major role in the aetiology of human insulin-dependent diabetes.

We were interested in the possible role of lymphokines in islet graft rejection as well as the specificity of the rejection reaction. In all these experiments, we used the DA rat as the recipient. After preparing the islets and hand-picking them, so that the numbers were exact, we implanted either 200 islets or 400 islets beneath the kidney capsule. The implanted islets were either syngeneic or allogeneic islets or a mixture of syngeneic and allogeneic islets. After implantation of the islets, the kidneys were removed after 3 weeks and the insulin extracted – remember that these islets were implanted under the kidney capsule. For example, a kidney with 200 syngeneic islets had 56×10^3 U insulin extracted from that organ, while double the number of syngeneic islets gave approximately double the amount of insulin. When 200 syngeneic and 200 allogeneic islets were implanted at opposite ends of the kidney, the same amount of insulin was extracted as from the kidney with 200 syngeneic islets only. Again, and this is the important experiment, where the islets were mixed intimately, we obtained the same amount of insulin as from a kidney with 200 syngeneic islets. And in the animals with only allogeneic islets, no insulin was extracted at all. What that implies is that although the allogeneic islets were rejected, presumably with the release of lymphokines, in particular IL-1 as well as IL-2, IL-4 and tumour necrosis factor, this did not produce any bystander damage of the syngeneic islets that were intimately mixed with the allogeneic islets. I should also point out that we have confirmed by immunoperoxidase histology using monoclonal antibodies to DA, PVG or Lewis

MHC antigens that in fact the only islets surviving in the mixture are the syngeneic ones.

These experiments have two implications. Firstly, they show the exquisite specificity of the rejection reaction in this model, which is not compatible with a delayed-type hypersensitivity type of mechanism of rejection. Secondly, and this is perhaps more important, they suggest that perhaps IL-1 does not produce damage to islets or β-cells. However, we must still demonstrate that IL-1 is actually released as part of the inflammatory reaction to the allogeneic islets.

I want now to go on to experiments in larger animals. Obviously, when these experiments first started 10–12 years ago, I thought that we would transplant islets clinically fairly soon. We have not, and although we are closer to clinical trials, we still have many problems to resolve. The first attempts employed a technique that Mirkovitch described in Lansanne which was really merely a crude pancreatic digest in which the pancreas was chopped up and digested with collagenase, and this pancreatic mush was then injected into the spleen. We could not inject it into the liver as the dogs all died of portal hypertension. Dogs that had a total pancreatectomy survived only about 6–7 days before they had to be put down with severe hyperglycaemia. Dogs that received an autotransplant survived for 2 months and were normoglycaemic during that time. However, they were really quite ill and were put down deliberately at 2 months. At autopsy they had abscesses within the spleen and extensive inflammatory reaction around the spleen. We therefore felt at the time (about 8 years ago) that this would not become a clinically applicable technique. However, it is interesting that the recipients of allografts, although never achieving normoglycaemia, did show some prolongation of survival over control animals.

The problem with the techniques for preparing islets that were so successful in rodents is that they are virtually useless in large mammals such as the dog, pig, primate or man. And then D. Gray, a Ph.D. student of mine at that time, developed a technique which was really quite successful. Basically it involved two major modifications to a variety of approaches used by ourselves and others: firstly the intraductal injection of collagenase warmed to 37°–39°C, and secondly after digestion of the pancreas in this way the islets were then separated either by filtration or on a multidensity gradient. The preparation that we obtained at the time was about 30% pure as judged histologically. It is crucial, we feel, that any comments about purity be based on histological assessment. Purity of 40% was the highest level that we attained, and in terms of numbers we obtained from about 600 to a maximum of 2500 islets per gram of pancreas. Since then we have continued our work on these preparations and have introduced a number of modifications so that we can now obtain about 2000–3000 islets per gram of pancreatic tissue with a purity of 50%–70%.

However, we are still not satisfied with the technique, for we cannot do this reproducibly. It is still an art rather than a science. And this is very worrying in terms of commencing clinical trials. Nextly, how did we know that these islets were alive? First we used fluorescine-diacetate and ethidium bromide, which very nicely discriminates between viable and non-viable islets. It is a crude but a quick screening technique for viability. These islets also produce insulin in vitro in response to an increase in glucose concentration in the medium, and finally they can be transplanted under the kidney capsule of a nude mouse or rat and show that these isolated human pancreatic islets survive, as judged by stain for insulin some weeks later, or the correction of streptozotocin-induced diabetes.

We then felt that we were approaching clinical trials, and about 3 years ago we decided to embark on primate experiments as a preclinical model using the cynomolgus monkey. There have been some disappointing features about these experiments. Because the cynomolgus monkey has a pancreas that microscopically and macroscopically resembles the human pancreas, it seemed a satisfactory animal in which to test the technique. We were also interested in autologous grafts to see how long these isolated islets would survive, and how they would behave physiologically. Implantation of autologous grafts into the spleen of the primate after total pancreatectomy results in normoglycaemia but not a marvellous response in some of the animals, at least in the first 3 days. However, when you take the spleen out at 6 weeks, the animals become diabetic again, confirming that it was the autologous reimplanted islets that were maintaining the animals normoglycemic.

We then changed to intraportal injection; because these are relatively pure islet preparations, there is no problem with portal hypertension after injection of these islets directly into the portal vein. I think it fairly crucial for clinical use that one employ as pure a preparation as possible. Following intraportal injection all animals initially became normoglycaemic, however the worrying thing is that failure occurred from 32 days to over 3 years in the last two animals, who both have just recently failed.

That is quite worrying, because although we would like to think that it is just a numbers game, namely that we did not implant enough pure islets, we cannot be sure of this. It may be somewhat more subtle, but we are obviously repeating all these experiments, using better yields and more pure preparations. Subsequent study of the successfully transplanted animals shows that these islets have a relatively good metabolic response. Their glucose tolerance test is abnormal, but not grossly so, and the response in terms of insulin secretion is qualitatively quite normal, although quantitatively only about one-quarter that of a normal animal. This in fact represents about the number of islets that have been transplanted, namely about one-quarter of the normal islet population. Finally, the only correlation that we found with long-term survival was the intravenous glucose tolerance test 6 weeks after transplantation. If it is a good response, then these are the animals that go on to the longest normoglycaemic survival. And this makes sense. They are probably the animals that have been given the most viable and functional islets. Survival did not correlate with islet numbers per se.

So where do we go in the future? I think we are almost ready for clinical trials. I would not like to pretend that success is just around the corner, and when we do start, the only thing that we are hoping to achieve and would regard as a success is to render someone insulin independent, even if for only 1 week. To be realistic, we may not hope for any more then this the first time round. I should point out that there have been a number of clinical trials of islet transplantation using impure preparations, and more recently pure preparations in St. Louis and in Miami, and although there has been some encouraging success in terms of the peptide levels being elevated after transplantation, no patient had been rendered insulin independent – which really must be the final yardstick of success. (Since this lecture was given, two patients in St. Louis have been rendered insulin independent after islet transplantation, one for 2 weeks and one for 2 months.)

Total Lymphoid Irradiation in Transplantation: Experimental Background and Results in 70 Patients*

J.A. Myburgh

Introduction

It is a great honour and privilege to participate in this symposium. The value of these symposia, held in incomparable surroundings, far transcends the mere exchange of views and information. They will be a lasting tribute and memorial to the insight, wisdom and humanity of Prof. Walter Brendel. A special tribute must also go to Mechthild Stein for her role in the impeccable organization of these unique meetings.

The purpose of this report is to give a brief overview of our studies in the use of total lymphoid irradiation (TLI) in transplantation, and to present an update of our experience with its use in clinical kidney transplantation.

Transplantation Tolerance

The production of transplantation tolerance remains the Holy Grail of clinical transplanters throughout the world. At an operational level it may be defined as the production of permanent survival of an allograft with normal function in a healthy recipient, without the need for permanent immunosuppressive drug therapy. At an immunological level it is usually held to be the induction of permanent and specific unresponsiveness of the recipient to the donor allograft. This state can readily be achieved in certain defined circumstances, such as the induction of neonatal tolerance in the mouse and enhancement of kidney grafts by donor-specific alloantibody in certain rat strains. The production of transplantation tolerance in large outbred animals, particularly primates, has been much more elusive. For more than two decades we have studied this phenomenon in the chacma baboon (*Papio ursinus*) using many immunological manipulations, some of which are very successful in rodent experiments [1–3]. Significant prolongation of kidney and liver allograft survival was attained, but with very rare exceptions nothing approximating true operational or immunological tolerance.

* This work was supported by the Council of the University of the Witwatersand and the South African Medical Research Council.

K. Messmer M. Stein (Eds.)
Pathways in Applied Immunology

The experiments of Slavin et al. [4] in the late 1970s, reporting the production of tolerance for skin and heart allografts in mice and rats by a modified method of TLI and donor bone marrow injection immediately raised the question with us whether this would once again be a purely rodent phenomenon, or whether it might be applicable to transplantation in outbred primates.

Experimental Studies in Primates

We have previously reported these studies extensively [5]. Briefly, pilot studies showed rapidly that TLI was uniquely powerful in kidney and liver transplantation in comparison with all our previous attempts at tolerance production in this species. However, it was also apparent that the dose of irradiation used in the rodent experiments was associated with unacceptable morbidity and mortality. Accordingly we embarked on many hundreds of experiments in which a series of variables of potential relevance to the safety and efficacy of TLI in baboons were studied systematically. These included variations in the irradiation regimen, including cumulative dose, frequency and doses of irradiation fractions, the field required to be irradiated, variable delays between completion of irradiation and transplantation – how long the window of tolerance induction would remain open – the necessity for donor bone marrow injection with the potential hazard of graft-versus-host disease, and the necessity for and role of adjuvant immunosuppressive drug therapy.

The optimal regimen which emerged from these studies consisted of a cumulative dose of 800 cGy, given in twice- or thrice-weekly fractions of 80 or 100 cGy each to a wide field. The irradiation field required was more extensive than the classical mantle-inverted Y field used in the treatment of patients with Hodgkin's disease and involved the entire torso with only the skull and extremities being shielded from irradiation.

This regimen produced tolerance in one-third of baboons, without the concomitant administration of any immunosuppressive drugs and without donor bone marrow injection. Indeed, donor bone marrow injection at the time of transplantation was counterproductive, and with rare exceptions tolerance production was abrogated by this manoeuvre.

The tolerance induced was both durable and specific. Tolerant baboons have survived in a perfectly healthy condition and with normal graft function for periods of more than 10 years after kidney or liver allotransplantation. None of more than 50 baboons with normal graft function 1 year after transplantation has subsequently rejected their grafts. The tolerance produced is specific in that these animals re-accept donor tissue, either kidney or skin, but reject grafts from unrelated third-party donors. The specifically tolerant state takes several months to develop. In the early months after transplantation the animals are either nonresponsive or hyporesponsive to allogeneic stimulation both in vivo or in vitro (mixed lymphocyte reactivity, MLR), but between 6 months and 1 year after transplantation the nonresponsiveness becomes purely donor-specific. A nonspecific, non-T suppressor cell phenomenon is demonstrable in the early, broadly hyporesponsive period after transplantation. This wanes and is no longer demonstrable by the end of 1 year. Attempts to demonstrate a specific CD8 suppressor T cell mechanism have been largely unsuccessful. Of practical significance was the demonstration that, in contrast to the situation in rodent experiments, tolerance could still be induced with delays of up to 3 months between completion of irradiation and transplantation, provided booster fractions of 80 or 100 cGy were given at 2–4 week intervals. The administration of a brief course (2–3 weeks) of conventional immunosuppressive drugs after transplantation was not synergistic with, or even additive to, the effect of pretransplant TLI. More dramatic was the demonstration of a counterproductive effect of additional fractions of TLI given after transplantation: true tolerance ensued in only a single animal so treated. In vitro studies demonstrated that the early post-transplant period of tolerance induction is one of intense immunological activity, as evidenced by the precipitous rise in the prevalence of activated T cells in the circulation. It is conceivable that injudicious immunological interference in this critical period, by irradiation or even immunosuppressive drug therapy, could counteract or even abrogate the process of tolerogenesis. This phenomenon has potentially disturbing implications for the use of TLI in the circumstances of human organ transplantation.

TLI in Clinical Kidney Transplantation

Following the promising and gratifying results obtained in baboons, TLI has been used in the preparation of an increasing number of patients undergoing renal transplantation.

Patients and TLI Regimen. Pretransplant TLI has been administered to 70 patients receiving kidney transplants. The irradiation regimen was the same as that found to be most effective in baboons, namely a cumulative dose of 800 cGy given as twice-weekly fractions of 80 cGy each to a wide field. Supra-diaphragmatic lung fields and the gonads were shielded. Booster fractions 1–4 at 2- to 4-week intervals of 80 cGy each were given if delays occurred before transplantation. The longest delay was 4.5 months. There were 54 cadaver donor (CD) transplants, of whom 11 received second or third transplants. All 16 related living donor (RLD) transplants were haploidentical. Twenty-eight of the recipients were children under 15 years.

Follow-Up. Follow up has ranged from 4 months to 8.5 years. Sixty patients have been at risk for over 1 year, 80% for over 2 years, and over one-half (37) have been followed up for over 3 years since transplantation.

One-Year Actual Survival. One-year actual patient survival is 95% (59/62). One-year actual graft survival for first CD transplants is 95% (37/39), and the figures for CD retransplants and RLD transplants are 80% (8/10) and 85% (11/13), respectively.

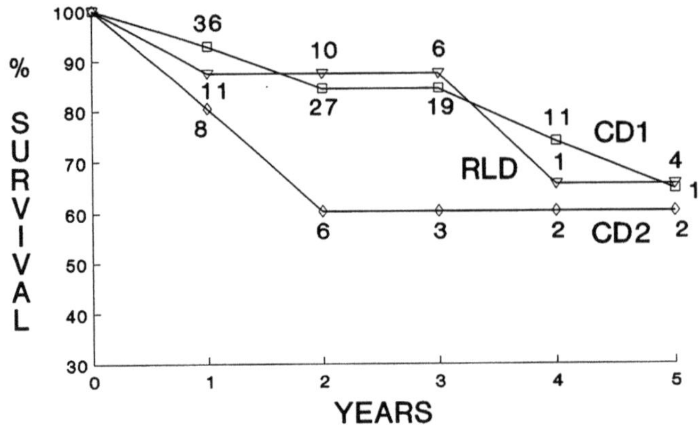

Fig. 1. Actuarial kidney graft survival of patients receiving TLI

Actuarial Graft Survival. The 1- to 5-year actuarial graft survival is depicted in Fig. 1. The figures for first CD transplants (CD1) are 93% and 65%, for RLD transplants 88% and 66%, and for CD retransplants (CD2) 80% and 60%.

Current Functional Graft Status. Fifty-two grafts (74%) are currently functional. Serum creatinine levels are normal in 22 patients (31%) and satisfactory (131 to 250 µmol/l) in 23 patients (33%). Graft function is poor, serum creatinine over 250 µmol/l, in 10 patients (14%).

Graft Rejection. Fourteen grafts (20%) have been lost by rejection, 7 of 43 (16%) of first CD transplants, 5 of 11 (45%) CD retransplants ($p = 0.05$) and 2 of 16 (13%) of RLD. There was no significant difference in graft loss between adult (8/42, 19%) and paediatric (6/28, 21%) recipients. Six of these patients have been retransplanted, and three are currently functional. Five grafts were lost in the 1st year after transplantation and five in the ensuing year.

Immunosuppressive Drug Therapy. All except four patients received cyclosporine (CsA) 6 mg/kg per day and prednisone 20 mg per day after transplantation. The dosages were tapered fairly rapidly; the mean current doses in the patients more than 1 year after transplantation are given in Table 1. Four of the 19 patients on CsA and prednisone receive alternate day steroid therapy, as do 1 of the 6 on azathioprine and prednisone, and 3 of the 17 on triple therapy with CsA, azathioprine and prednisone. Five patients receive prednisone only at a mean dose of 3.7 mg per day.

In ten patients CsA was tapered and withdrawn entirely 43–81 days after transplantation. A gradual rise in serum creatinine led to reintroduction of either CsA or azathioprine 5.3 ± 3.0 months later in eight patients, in one of whom CsA was again withdrawn later. Currently, 63 ± 8 months after transplantation, five of these ten patients have normal serum creatinine levels.

Table 1. Mean daily doses of immunosuppressive drugs in patients over 1 year after transplantation

	n	CsA (mg/kg per day)	Aza (mg per day)	Pred (mg per day)
CsA + Pred	15	3.3	–	8.5
CsA + Pred q.o.d.	4	4.3	–	7.5
Aza + Pred	5	–	68	9.2
Aza + Pred q.o.d.	1	–	50	12.5
CsA + Aza + Pred	14	3.9	48	9.5
CsA + Aza + Pred q.o.d.	3	1.5	33	10.0
Pred alone	5	–	–	3.7

CsA, Cyclosporine; Aza, azathioprine; Pred, prednisone.

Three patients still receive prednisone only, 2, 2 and 4 mg per day. In the remaining five patients, four grafts have been lost, and one has chronic rejection, with a serum creatinine of 343 µmol/l 56 months after transplantation.

In four further patients, CsA was withdrawn at a mean of 21 months after transplantation. CsA had to be restarted in one patient at 2 years, and this patient has chronic rejection, with a serum creatinine of 521 µmol/l at 49 months. Two patients retain normal graft function on prednisone only, 5 and 7.5 mg per day, 3.5 years after transplantation. The final patient had all immunosuppression withdrawn 3 years after transplantation. She retained normal function for a further 5.5 years when the serum creatinine rose slowly but progressively. Graft histology showed dense interstitial fibrosis only, with no features of cellular or humoral rejection. The nature of this process remains obscure, as she was shown to be specially nonreactive to donor-HLA matched lymphocytes in a cell mediated lymphocytotoxicity assay [6]. She has been retransplanted successfully.

The major identifiable factor in the patients who lost their grafts after CsA withdrawal was presensitization. The sera of all four patients reacted with 33% to 50% of a panel of lymphocytes, and three of the four had positive ^{51}Cr release cross-matches against donor lymphocytes, dye exclusion cross matches having been negative.

Deaths. Six patients died from transplant-related causes, four of these with functioning grafts. Two patients died of malignant conditions, one with myeloid leukaemia and one with carcinoma of the bladder and original kidney. There were three deaths from infection. Cytomegalovirus (CMV) was implicated in all three, and two patients had superimposed bacterial infection. There was one death from suicide after rejection of a graft.

Infections. Twenty-three patients experienced bacterial infections. The majority were treated easily, but six patients developed gram-negative septicaemia. Two patients experienced legionellosis, both of whom responded to treatment, and one patient contracted tuberculous pericarditis. Fourteen patients contracted viral infections. CMV was the commonest, occurring in 11 patients, of whom 3 died. There were two patients with herpes zoster and a single patient with an unidentified viral meningitis. There were two patients with fungal infections, cryptococcal meningitis and torulopsis meningitis; both patients survived.

Immunological Studies on the Mechanisms of TLI

Donor-specific unresponsiveness is generally held to be the in vitro correlate of in vivo transplantation tolerance. We have previously reported that this expected pattern was seen in 9 of 16 operationally tolerant baboons studied 1.5–4 years after TLI and kidney transplantation [7]. However, two other patterns of MLR were seen as well: in one-third of tolerant animals studied there was a broad nonspecific hyporesponsiveness to both donor and third-party lymphocytes, and, significantly, in two baboons studied 4 years after transplantation their lymphocytes responded vigorously to both donor and third-party lymphocytes despite full operational tolerance. We have seen the same three patterns in patients following TLI and kidney transplantation.

As mentioned previously, a nonspecific non-T suppressor cell phenomenon can be demonstrated regularly in the early months after TLI and transplantation, but we have found no convincing evidence for CD8-mediated T cell suppression to explain specific unresponsiveness. In animals with donor-specific unresponsiveness in MLR, CD8 depletion of responder lymphocytes fails to restore responsiveness to donor lymphocytes in MLR. MLR responsiveness is, however, partially restored by exogenous recombinant interleukin 2 (rIL-2). TLI in doses of more than 500 cGy suppresses the induction by concanavalin A of IL-2 receptors (submitted for publication). These findings suggest that inhibition of IL-2 production is an important mechanism of action of TLI. The mechanism of evolution of specificity of tolerance remains to be elucidated.

We have also reported [4] the presence of a potent plasma suppressor factor in the serum of operationally tolerant baboons. The factor has been purified into two fractions, having molecular weights of 750 and 145 kDa, respectively. Preliminary results indicate that this factor may be an IgG, and that the higher molecular weight may be due to forced molecular aggregation (submitted for publication).

Conclusions

TLI offers encouraging possibilities for the realization of the ultimate goal of clinical transplantation, namely the production of specific tolerance. It can produce durable and specific tolerance in the subhuman primate. Results in clinical human transplantation are gratifying in terms of patient and graft survival. It permits the use of very low doses of concomitant immunosuppressive drugs with a commensurately low prevalence of serious side effects. Further experience should reveal how frequently a truly tolerant state can be achieved in the human situation, given the existence of unavoidable complicating circumstances, particularly the impact of presensitization. Prediction of the evolution of true tolerance in the individual is currently impossible. A better understanding of the underlying mechanisms involved should allow a more rational manipulation of existing tactics and strategies, with a more predictable outcome as the consequence.

Acknowledgements. The collaboration and contributions of many colleagues and assistants is gratefully acknowledged.

References

1. Myburgh JA, Smit JA (1972) Passive and active enhancement in baboon liver allografting. Transplantation 14:227–235
2. Myburgh JA, Smit JA (1975) Prolongation of liver allograft survival by donor-specific soluble transplantation antigens and antigen-antibody complexes. Transplantation 19: 64–71
3. Myburgh JA, Smit JA (1979) Autoantiidiotypic immunization and transplantation tolerance in the primate. Transplant Proc 11:923–927
4. Slavin S, Strober S, Fuks Z, Kaplan HS (1977) Induction of specific transplantation tolerance using fractionated total lymphoid irradiation in mice: long term survival of allogeneic bone marrow and skin grafts. J Exp Med 146:34–48
5. Myburgh JA, Smit JA, Meyers AM, Botha JR, Browde S, Thomson PD (1986) Total lymphoid irradiation in transplantation. World J Surg 10:369–380
6. Strober S, Dhillon M, Schubert M, Holm B, Engleman E, Benike C, Hoppe R, Sibley R, Myburgh JA, Collins G, Levin B (1989) Acquired immune tolerance to cadaveric renal allografts. A study of three patients treated with total lymphoid irradiation. N Engl J Med 321:28–33
7. Myburgh JA, Meyers AM, Thomson PD, Botha JR, Margolius L, Lakier R, Smit JA, Stark JH, Gray C (1989) Total lymphoid irradiation–current status. Transplant Proc 21:826–828

Developments in Bone Marrow Transplantation

D.W. van Bekkum

Introduction

It is an honour as well a pleasure to participate in this 20th anniversary of the famous round table meetings of Prof Walter Brendel. I have had the privilege of being present at most of these conferences, and this has given me the opportunity to interact with a variety of outstanding scientists. It has also been a source of inspiration and has added to my education. At these meetings we have also been exposed to scores of associates of Walter's, and over the years we have witnessed their development under his guidance. I have always admired Walter's aptitude for selecting these researchers, who have contributed so much in achieving the ideal blend of sports, pleasure and serious scientific exchange which have been the outstanding features of these meetings.

My contributions at these meetings have usually concerned aspects of bone marrow transplantation, and the present one is no exeption. The first part deals with a development – T cell depletion of the graft – which has made bone marrow grafting a much less hazardous treatment, thereby opening the way for a larger variety of clinical applications. In the second part I discuss the use of bone marrow transplants in the treatment of lysosomal enzyme deficiencies, as well as new experimental data on the effect of bone marrow grafting on autoimmune diseases, in particular experimental arthritis.

Prevention of Graft-Versus-Host Disease by T Cell Depletion

The natural incidence of graft-versus-host disease (GvHD) in patients following MHC-matched bone marrow transplantation is not precisely known since they are nearly all treated with immunosuppressive agents after transplantation. Methotrexate was employed during the early days of clinical bone marrow transplantation; presently cyclosporin A alone or in combination with methotrexate is given. In the methotrexate series the incidence of GvHD was 50%, and in one half of the patients who suffered from GvHD this complication was fatal.

In outbred animals the influence of MHC matching has been studied without immunosuppressive treatment. In rhesus monkeys transplanted with fully MHC-matched marrow the incidence of severe, fatal GvHD was 90%.

K. Messmer M. Stein (Eds.)
Pathways in Applied Immunology

Transplantation of MHC-matched sibling marrow in dogs results in about 50% GvHD, but its severity is less, and the disease is usually not fatal. In H2-identical mouse strain combinations, GvHD does not usually complicate bone marrow transplantation. These differences between species largely disappear when MHC-mismatched marrow is grafted. Dogs and monkeys as well as rodents (mice and rats) develop severe and fatal GvHD, but in rodents the disease shows a different pattern in that it runs a delayed rather than an acute course. The difference in responses between species can be fully explained by differences in the proportion of T lymphocytes in their bone marrow: a high proportion, up to 30% in primates, only 1%–2% in rodents, and an intermediate level in bone marrow of dogs. In the early days of bone marrow transplantation research, we were able to demonstrate that the cells responsible for the induction of GvHD are lymphocytes, later defined as the subclass of T lymphocytes. In mice it was shown that by adding lymphoid cells to the marrow graft, GvHD can be evoked in donor recipient combinations where it does not normally occur. Furthermore, the GvHD can be aggravated by increasing the number of lymphoid cells [16]. In this way the acute form of GvHD as seen in patients and monkeys can be reproduced in the mouse. The consequence of this observation is that GvHD would be prevented by removal or selective inactivation of lymphocytes in the graft prior to transplantation. In 1968 we reported a method involving centrifugation on density gradients for separating the lymphocytes responsible for the GvHD from the cells required to repopulate the immunohaemopoietic system of lethally irradiated mice [3]. Subsequently, adaptation of these methods to monkey, dog and human marrow fully established that T lymphocyte depletion of the bone marrow graft is highly effective in preventing GvHD [19].

Inactivation of T lymphocytes with monoclonal antibodies alone or in combination with physical methods of separation is now also employed. The advantage of lymphocyte depletion was first confirmed in the clinic in patients with severe combined immune deficiency [17]. Recent combined data from a large number of clinical centres shows the results obtained with T cell depleted marrow from HLA-mismatched donors to be equivalent to those of HLA-identical grafts [4]. In recent years lymphocyte depletion has been used in the treatment of leukaemia and aplastic anaemia by an increasing number of bone marrow transplant teams. To evaluate the efficacy of GvHD prevention, monitoring of the number of T lymphocytes before and after depletion is required. In addition, it is vital to assure that the depletion technique does not decrease the number of viable haemopoietic stem cells in the graft.

Most investigations have reported a sharp decrease of the incidence of GvHD or even its complete absence following the grafting of T lymphocyte depleted marrow from MHC-matched donors. However, several transplant teams have also encountered two problems with this approach. The first is an increased incidence of take failures; the second, relating only to leukaemic patients, is an increased incidence of leukaemic relapses. Our experimental data predict that both complications can be overcome by intensification of the conditioning regimen, provided that this is accomplished by cytotoxic agents.

Rejection of T Cell Depleted Grafts

The notion that lymphocyte-depleted bone marrow is more prone to rejection was predicted by observations in a variety of experimental systems, which suggested a reciprocal interference between HvG and GvH reactions [21]. More than 30 years ago it was shown that recipient-type lymphoid cells are capable of rejecting rat bone marrow grafted into lethally irradiated mice [18]. At that time it was postulated that rejection was mediated via production of antibodies against the graft, but more recent evidence suggests that the effector cells of the HvG reaction are T lymphocytes. The strongest evidence for this assumption was provided by Bau and Thierfelder [1], who were able to obtain stable mouse-to-rat bone marrow chimeras by conditioning with supralethal total body irradiation (TBI) and pretreatment of the recipients with a high dose regimen of anti-rat lymphocyte serum.

A significant incidence of take failures was observed in monkeys receiving T cell depleted mismatched bone marrow, following conditioning with a dose of TBI that permitted a 100% take of whole bone marrow grafts [19]. No take failures were seen with purified stem cell grafts from MHC-matched siblings in rhesus monkeys [22–24]. In contrast, in the dog a high rate of take failure was observed with purified stem cell grafts from DLA-identical sibling donors. The only difference between the dog and the monkey model was that in the dog a slightly lower dose of TBI (7.5 Gy) was employed than in the monkey (8.8 Gy), these respective doses being the highest tolerated without the risk of complications from intestinal damage. When TBI is employed in the conditioning of human patients, the comparable dose is usually 8.5 Gy at a dose rate of 0.2 Gy per min, which is biologically equivalent to 10 Gy at 0.04 Gy per min. From these data it is difficult to predict the outcome of purified stem cell grafts in human patients. It is not known whether they will resemble dogs or monkeys in their responses. Therefore, in dogs conditioning regimens were studied which would allow reproducible takes of purified stem cell grafts in DLA-identical siblings. Two equal fractions of 4.5 Gy each, delivered at an interval of 24 h, served this purpose and were well tolerated [25,26]. In monkeys, two fractions of 7 Gy each (separated by 72 h) permitted the take of purified stem cell fractions form MHC-mismatched unrelated donors [22,24].

These results were used in Rotterdam in the treatment of leukaemic patients receiving MHC-matched marrow; Löwenberg et al. [11] employed a conditioning of cytosine arabinoside, cyclophosphamide, and two fractions of 5 Gy TBI separated by 72 or 24 h (lungs shielded to a maximum dose of 8.5 Gy). In all of the 17 evaluable patients engraftment occurred [5]. Another group has experienced no adverse effects of T cell depletion on take rate when a single-dose TBI was employed for conditioning in combination with cyclophosphamide [14]. One British transplant team has reported an alarmingly high rate of graft failures following the introduction of T cell depleted bone marrow grafting. When their conditioning regimen – consisting of cyclophosphamide and six fractions of 2 Gy TBI over 3 days – was intensified by one additional fraction of 2 Gy, the graft failure rate fell to 1 out of 27 transplants [2].

These clinical data agree very well with observations in mice, where we titrated the number of recipient-type spleen cells required for the rejection of allogeneic bone marrow grafts by lethally irradiated recipients. We found that the trajectory between 50% and 100% graft take comprises less than log_1 of host-type T lymphocytes. An additional log_1 of T lymphocytes requires between 1.5 and 2.5 Gy depending on whether the TBI is delivered as a single dose or in fractions. More experimental data are clearly needed to sort out the quantitative aspects of graft rejections in relation to the number of residual T lymphocytes, but so far there is a fair degree of consistency between the results obtained with laboratory animals and the clinical experience.

Increased Relapse Rate Following the Grafting of T Cell Depleted Marrow

The second problem in the use of a T cell depleted bone marrow graft is that in the absence of GvHD – irrespective of how this is achieved – the incidence of leukaemic relapse is greater. Weiden et al. [27] in a large series of patients grafted with MHC-matched sibling marrow, reported a relapse rate of 10% in patients who had experienced both acute and chronic GvHD, against a relapse rate of 60% in those who showed no signs of GvHD. It has not yet been established whether the anti-leukaemic effect associated with GvHD is due merely to alloreactivity or to a specific graft anti-leukaemia effect, but for the following argument this is not important. If it is assumed that a relapse may occur as the result of the survival of a single leukaemic clonogenic cell, the theoretical probability of leukaemia cure can be calculated with either Poisson statistics or binominal probability distribution functions. In this way it can be calculated that the difference between 90% cure (10% relapse) and 40% cure (60% relapse) corresponds with a difference of log_1 leukaemic cell kill [15].

Exact data on the radiosensitivity of leukaemic cells are not available, but it may be safely assumed that it is within the range of that estimated for hemopoietic stem cells (HSC) and lymphocytes [13]. It follows that the additional dose of radiation needed to reduce a 60% relapse to 10% is approximately 2 Gy. This additional radiation needed to assure full engraftment with T cell depleted marrow also inactivates sufficient residual leukaemic cells to compensate for the loss of the anti-leukaemic effect of GvHD. Similar predictions can be made for extra immunosuppressive conditioning with chemical agents, provided that these drugs are equally effective against lymphocytes as against leukaemic cells. If, however, the extra immunosuppression is effected with specific anti-lymphocytic antibodies, additional anti-leukaemic treatment should be given either as conditioning or as consolidation therapy after the bone marrow transplantation. A similar restriction applies to the use of total lymphoid irradiation as an additional immunosuppression, since a large proportion of the bone marrow that is likely to contain residual leukaemic cells is not irradiated with this technique.

Bone Marrow Transplantation in the Treatment of Hereditary Enzyme Deficiencies

A variety of hereditary deseases, among them the so-called lysosomal storage diseases, is due to the lack of a single enzyme. Systemic administration of the missing enzyme, if at all beneficial, has so far been found to be only temporarily effective. In those cases when the deficient enzyme is produced by haemopoietic cells, grafting of normal bone marrow may provide a continuous supply of that enzyme [6]. Even mixed chimerism, provided it were stable, would be expected to cure these patients. The feasibility of this approach has been demonstrated in several animal models as well as in patients with various diseases of this category [10].

However, in many of these syndromes the central nervous system is progressively damaged as a result of accumulation of storage products. There is no convincing evidence yet that a successful marrow graft causes regression of the pathological processes in the CNS. Experiments with mice suffering from hereditary galactosyl-ceramidase deficiency (the analogue of Krabbe's disease in man), showed that in contrast to other tissues, the enzyme levels in the CNS are not restored following bone marrow transplantation [7–9]. Although the level of the toxic product psychosine, which is involved in the demyelination of the nerve fibres in this disease, fell significantly in the grafted animals, the beneficial effect apparently occurred too late for adequate recovery or prevention of the lesions. Hoogerbrugge et al. [8] demonstrated the presense of donor derived macrophages in the CNS of galactosyl-ceramidase deficient mice following bone marrow transplantation, but their cells seem to penetrate the CNS only as a result of advanced tissue damage. The main challenge for the future therefore seems to devise means which allow better and earlier penetration of donor-derived cells or of the enzyme released by these cells into the nervous tissues. Another requirement is the design of less toxic conditioning regimens for these patients, allowing partial or complete engraftment without undue risks of aplasia or GvHD. With the current agents this has not been achieved, so that new immunosuppressive agents and preferably novel approaches must be explored.

Treatment of Autoimmune Diseases with Bone Marrow Grafts

An increasing number of serious, crippling diseases, some of them life threatening, have been attributed to autoimmunity. Among them are juvenile type 1 or insulin-dependent diabetes, rheumatoid arthritis, lupus erythematosus and multiple sclerosis. The pathogenesis of these diseases is associated with a genetic predisposition in conjunction with a variety of external initiating factors. Depending on the disease, humoral or cellular autoimmune reactions, or a combination of both are involved. The end result is the destruction of a specific target tissue. Various animal models of autoimmune disease have become available, among them the NOD mouse and the BB rat as models for type 1 diabetes; arthritis induced in rats and mice by immunization with adjuvant, collagen I or streptococcus A cell walls and experimental allergic encephalitis as a model for multiple sclerosis. Only certain strains of rodents are susceptible to the induction of autoimmune disease. In several of these models, evidence has accumulated suggesting that the genetic susceptibility resides in the haemopoietic stem cell.

Transplantation of bone marrow between susceptible and non-susceptible strains has demonstrated that susceptibility is determined by the origin of the bone marrow in residence. The spontaneously arising diabetes in BB rats and in NOD mice can be prevented by replacing the haemopoietic system cells by that from a non-diabetic donor strain, provided that it is performed before the Langerhans' islets are destroyed [12,28]. In our institute, complete and lasting remission of adjuvant arthritis in rats resulted from high-dose TBI and grafting of allogeneic bone marrow from non-susceptible donors, provided that the animals were treated during the acute stage of the disease, i.e. 4 weeks after the administration of adjuvant [20].

Rats treated during the chronic stage of the arthritis showed complete regression of the inflammatory reactions, but the existing bone deformations were not affected. Surprisingly, the acute arthritis could also be cured by transplantation of syngeneic bone marrow derived from healthy animals and even by autologous bone marrow [29]. Clearly, these observations may provide new opportunities for studying the pathogenesis of autoimmune diseases. Furthermore, these findings will lead to the exploration of bone marrow transplantation for the treatment of other experimental autoimmune diseases. It is too early to advocate that bone marrow transplantation be used to treat severe autoimmune diseases in patients. The major obstacle for clinical experiments is the high toxicity of the procedure, as has been pointed out in the case of enzyme deficiency diseases. The development of new, more tolerable methods for replacing the haematopoietic system deserves high priority.

The experimental models of autoimmune disease may be employed to establish whether mixed chimerism or temporary replacement of the haemopoietic system is similarly effective as complete and permament replacement.

References

1. Bau J, Tierfelder S (1973) Antilymphocytic antibodies and marrow transplantation. I. Transplantation 15:564
2. Burnett AK, Hann IM, Robertson AG, Alcorn M, Gibson B, McVicar I, Niven L, Mackinnon S, Hambley H, Morrison A, Todd A (1988) Prevention of graft-versus-host disease by ex vivo T cell depletion: reduction in graft failure with augmented total body irradiation. Leukemia 2:300
3. Dicke KA, Van Hooft JIM, Van Bekkum DW (1968) The selective elimination of immunologically competent cells from bone marrow and lymphatic cell mixtures. II. Transplantation 6:562
4. Fischer A, Griscelli C, Friedrich W, Kubanek B, Levinsky R, Morgan G, Vossen J, Wagemaker G, Landais P (1986) Bone-marrow transplantation for immunodeficiencies and osteopetrosis. European survey (1986). Lancet ii:1080–1084

5. Hagenbeek A, Löwenberg B (1987) Prevention of graft-versus-host disease without take failures after allogeneic bone marrow transplantation. In: Bone marrow transplantation Vol. 2, Suppl 1, p 141
6. Hobbs JR (1981) Bone marrow transplantation for inborn errors. Lancet ii: 735–739
7. Hoogerbrugge PM, Poorthuis BJHM, Romme AE, Van de Kamp JJP, Wagemaker G, Van Bekkum DW (1988) Effect of bone marrow transplantation on enzyme levels and clinical course in the neurologically affected twitcher mouse. J Clin Invest 81:1790–1794
8. Hoogerbrugge PM, Suzuki KSK, Poorthuis BJHM, Kobayashi T, Wagemaker G, Van Bekkum DW (1988) Donor-derived cells in the central nervous system of twitcher mice after bone marrow transplantation. Science 239:1035–1038
9. Ichiokaf T, Kishimoto Y, Brennan S, Santos GW, Yeager AM (1987) Hematopoietic cell transplantation in murine globoid cell leukodystrophy (the twitcher mouse): effects on levels of galactosylceramidase, psychosine, and galactocerebrosides. Proc Natl Acad Sci USA 84:4259–4263
10. Krivit W, Paul NW (eds) (1986) Bone marrow transplantation for treatment of lysosomal storage diseases. Liss, New York
11. Löwenberg B, Wagemaker G, Van Bekkum DW, Sizoo W, Sintnicolaas K, Hendriks WDH, Hagenbeek A (1986) Graft-versus-host disease following transplantation of 'one log' versus 'two log' T-lymphocyte-depleted bone marrow from HLA-identical donors. Bone Marrow Transplant 1:133
12. Naji A, Silvers WK, Bellgrau D, Bakker CF (1981) Spontaneous diabetes in rats. Science 213:1390–1392
13. O'Donoghue JA, Wheldon TE, Gregor A (1987) The implications of in-vitro radiation-survival curves for the optimal scheduling of total-body irradiation with bone marrow rescue in the treatment of leukaemia. Br J Radiol 60:279
14. Prentice HG, Blacklock HA, Janossy G, Gilmore MJML, Price-Jones L, Tidman N, Trejdosiewicz LK, Skeggs DBL, Panjwani D, Ball S, Graphakos S, Patterson J, Ivory K, Hoffbrandt AV (1984) Depletion of T lymphocytes in donor marrow prevents significant graft-versus-host disease in matched allogeneic leukaemic marrow transplant recipients. Lancet i:472–476
15. Schultz FW, Vriesendorp HM, Hagenbeek A (1991) On the quantitative role of graft-versus-host disease in decreasing the leukemia relapse rate after allogeneic bone marrow transplantation. (to be published)
16. Van Bekkum DW (1964) The selective elimination of immunological cells from bone marrow and lymphatic cell markers. I. Transplantation 2:293
17. Van Bekkum DW, Dicke KA (1972) Treatment of immune deficiency disease with bone marrow stem cell concentrates. In: Porter R, Knight J (eds) Ontogeny of acquired immunity. CIBA Foundation Symposium, London, 23–25 Nov 1971, Elsevier, Amsterdam, p 233
18. Van Bekkum DW, Vos O (1957) Immunological aspects of homo- and heterologous bone marrow transplantation in irradiated animals. J Cell Comp Physiol 50[suppl 1]: 139
19. Van Bekkum DW, Wagemaker G, Vriesendorp HM (1979) Mechanisms and avoidance of graft-versus-host disease. Transplant Proc 11:189
20. Van Bekkum DW, Kinwel-Bohré EPM, Houben PFJ, Knaan-Shanzer S (1989) Regression of adjuvant-induced arthritis in rats following bone marrow transplantation. Proc Natl Acad Sci USA 86:1090–1094
21. Vriesendorp HM (1985) Engraftment of hemopoietic cells. In: Van Bekkum DW (ed) Bone marrow transplantation. Dekker, New York, p 73 (chap 2)
22. Wagemaker G (1986) Problem and prospects of histocompatible bone marrow transplantation studies in rhesus monkeys. In: Hagenbeek A, Löwenberg B (eds) Minimum residual disease in acute leukemia. Nijhoff, Boston, pp 363–373
23. Wagemaker G, Vriesendorp HM, Van Bekkum DW (1981) Successful bone marrow transplantation across major histocompatibility barriers in rhesus monkeys. Transplan Proc 13:875

24. Wagemaker G, Heidt PJ, Merchav S, Van Bekkum DW (1982) Abrogation of histocompatibility barriers to bone marrow transplantation in rhesus monkeys. In:Baum JS, Ledney GD, Thierfelder S (eds) Experimental hematology today. Karger, Basel, p 111
25. Walma EP, Wagemaker G (1981) Fractionated total body irradiation as conditioning for bone marrow transplantation in dogs and monkeys. Meeting of the Netherlands Radiobiological Society and Netherlands Society for Radiotherapy, Leiden, October 10. Int J Radiat Biol 41:63 (abstr)
26. Walma EP, Vriesendorp HM, Zurcher C, Van Bekkum DW (1987) Engraftment of stem-cell-enriched bone marrow fractions in MHC-identical dogs after fractionated total-body irradiation. Transplantation 43:818
27. Weiden PL, Flournoy N, Thomas ED (1979) Antileukemic effect of graft-versus-host disease in human recipients of allogeneic marrow grafts. N Engl J Med 300:1068
28. Yasumizu R, Sugiura K, Iwai H et al. (1987) Treatment of type 1 diabetes mellitus in non-obese diabetic mice by transplantation of allogeneic bone marrow and pancreatic tissue. Proc Natl Acad Sci USA 84:6555–6557
29. Knaan-Shanzer S, Houben PFJ, Kinwel-Bohré EPM, van Bekkum DW (1991) Remission induction of adjuvant arthritis in rats by total body irradiation and antologous bone marrow transplantation. Bone Marrow Transplant (to be published)

Micrometastasis of Epithelial Tumors: Signum Mortis for the Patient or the Achilles' Heel of Cancer?

G. Riethmüller, K. Pantel, I. Funke, and G. Schlimok

Remembering Walter Brendel's long-standing enthusiasm for tumor immunology, we have chosen a topic dealing with our most recent approach to identifying and defining micrometastatic cancer cells in bone marrow. When we met Walter for the first time in the mid-1960s, he was fascinated, as were many of us, by the idea that tumors might follow the same rules of acceptance and rejection as allografts. At that time we were not fully aware of the common history of tumor immunology and transplantation immunology, which had cross-fertilized each other already for some time. We are fortunate that Avrion Mitchison is with us here as witness of the early days when Gorer in the search of tumor rejection antigens discovered the major histocompatibility complex of mice.

The mid-1960s was also the heyday of antilymphocyte serum, an intriguing new concept being spread by Rudolf Pichlmayr through Brendel's underground bunker in Munich. Antilymphocyte serum was then thought to be the ideal tool to prove or disprove the basic tenet of immune surveillance theory that developing tumors were continuously eliminated by the vigilant immune system. And, of course, we all thought that T lymphocytes were the main candidates policing the organism for malignant cells with aberrant behavior. Who would have thought that T lymphocytes were so much more concerned about the cells' interior than about the cell membrane! Indirectly, transplantation immunology helped to redirect tumor immunology when, after many years of routine kidney transplantation, it became clear that chronically immunosuppressed patients with kidney grafts did not develop the most common types of cancers more often than controls.

With the demise of T cells as main agents of immune surveillance the era of natural killer cells as the natural panacea of cancer commenced. Yet, up to the present day, the nature and the exact physiological role of this chameleon cell has remained enigmatic. With the advent of monoclonal antibody technology and molecular cell biology a completely new approach to tumor antigens and tumor markers was possible. A spate of data has appeared in recent years demonstrating the qualitative and quantitative abnormalities in tumor cells. Yet, a clear demonstration of rejection-inducing antigens in spontaneously arising tumors is still lacking. A major breakthrough was the identification of a gene coding for a tumor-specific transplantation antigen. Using tumor-specific cytotoxic T cell clones, gene transfection led to the identifica-

K. Messmer M. Stein (Eds.)
Pathways in Applied Immunology

tion of peptides derived from a major intracellular protein which induced tumor rejection in conjunction with the appropriate MHC antigens [1].

From the new insights into tumor-associated antigens defined by monoclonal antibodies and from new data on possible cellular rejection mechanisms, various therapeutic strategies were developed and tested in clinical trials. These trials benefited enormously from the availability of recombinant lymphokines and cytokines such as interleukin 2 and colony-stimulation factor. A major drawback in most of these immunotherapeutic trials was their application to patients in far advanced stages of disease. Thus it came as no surprise that adoptive or passive immunotherapy trials involving, for example, the transfer of lymphokine-activated killer cells or monoclonal antibody infusion were rather ineffective vis-à-vis a bulky tumor mass. On the other hand, adjuvant therapies carried out in the absence of direct proof of residual cancer were fraught with uncertainties and with the requirement of extended observation times.

Considering the fact that the prinicipal cause of death from cancer can be reduced to the early dissemination of metastatic cells from the primary tumor, it is obvious to apply new immunological techniques to detect disseminated tumor cells at the time of surgery of the primary tumor when the residual tumor burden is minimal, and individual life expectancy is greatest. To identify and isolate disseminated tumor cells at such an early stage of metastasis amounts to the proverbial search for a needle in a haystack. So far, tumor marker expression has been found to be heterogeneous in primary tumors, i.e., in searching for individual cells one must anticipate the existence of cells negative for a particular tumor marker. Therefore, from the outset of our studies we attempted to define tumor cells by markers not affected by the inherent genetic instability of cancer.

Cytoskeleton constituents are extremely reliable labels for the histogenetic origin of malignant cells [2]. Of the various cytoskeleton components cytokeratins are characteristic for all cells derived from epithelia. Among the large family of cytokeratin proteins the peptide number 18 proved to be restricted to cells of simple epithelia and to tumors derived from this tissue. Thus, cytokeratin 18 was a promising candidate to identify tumor cells derived from mammary and colorectal cancers. As there was a highly specific monoclonal antibody (CK 2) against cytokeratin 18 available, this marker offered itself for the detection of individual tumor cells in the bone marrow [3]. Using as a sensitive immunoenzymatic technique the alkaline phosphatase antialkaline phosphatase procedure, it became possible to detect tumor cells with a frequency of 10^{-5} nucleated bone marrow cells. Applying a monoclonal antibody against the CD45 leukocyte common antigen as second marker in a sensitive double-staining technique, we could demonstrate that the markers were mutually exclusive on bone marrow cells: thus on a large panel of bone marrow donors we never encountered $CD45^{+}$ cytokeratin 18^{+} cells [4]. Moreover, analysis of more than 100 patients without malignant disease demonstrated that epithelial cells, as defined by the cytokeratin 18 peptide, were absent from the bone marrow.

Having thus established the sensitivity and specificity of the method, we applied it to the diagnosis of disseminated tumor cells in more than 500 patients with mammary, gastric, or colorectal cancer. The incidence of micrometastatic cells in bone marrow was positively correlated with the conventional risk parameters of tumor extent. For mammary cancer, the incidence of cytokeratin-positive cells in marrow increased from about 23% in lymph node negative patients with small primary tumors to about 70% in patients with clinically manifest distant metastasis [5]. Also in colorectal cancer the incidence of epithelial tumor cells in bone marrow was strictly correlated with the Dukes stages of tumor progression [6]. In gastric cancer the incidence could be correlated to the histological type [7]. Thus, bone marrow epithelial cells were more frequently found in patients with the less differentiated invasive type of gastric cancer.

As to the clinical significance of the findings, it could be demonstrated that the relapse free survival of patients with micrometastatic cancer cells in bone marrow was significantly shorter than in patients of same clinical stage without marrow micrometastasis [7]. For patients with colorectal cancer this finding came as a surprise since the skeleton is not a preferred site of metastasis in this type of cancer [6]. A retrospective analysis of the clinical course of the disease in patients presenting with bone marrow micrometastasis revealed that the majority of relapses occurred in the liver, as expected. Therefore the presence of micrometastatic cells may be taken as a general indicator for the metastatic potential of an individual tumor. Furthermore, it may well be that the dynamics of metastatic disease in colorectal cancer is such that patients die of complications due to liver metastasis before they clinically exhibit bone metastasis. This hypothesis is supported by autopsy findings in patients with colorectal cancer obtained by conventional histology of bone marrow showing an unexpected high incidence of metastasis [8].

In search of an explanation for the variable clinical course of patients with marrow micrometastasis, we were interested to determine whether additional characteristics could be identified on the individual tumor cells. For this purpose we used several monoclonal antibodies against antigens associated with cell proliferation, including the epidermal growth factor receptor, transferrin receptor, nuclear antigen Ki67, and antibodies against oncogene products such as the mutated ras protein and the HER2/neu protein. This approach yielded quite variable results demonstrating a rather heterogenous expression of proliferation-associated markers on individual tumor cells [9].

A major goal of our studies was to develop this system into a monitoring technique allowing the assessment of adjuvant therapies which are still administered in a blinded fashion. It would be a great advantage to have a monitoring technique enabling the therapist directly to assess the efficacy of the various therapeutic strategies on residual and so far invisible tumor cells. As the technique can still be refined, and as bone marrow is an easily accessible tissue compartment, the prospects of such a monitoring system are quite promising. However, the technical requirements and the exigencies for specificity relying on the correct identification of very few cells are enormous. Thus the use of monoclonal antibodies reacting with antigens, for example, the epithelial (EMA) membrane antigen which is also present on hemopoietic cells, is detrimental for this approach of tumor cell detection [10]. Furthermore a multivariate analysis is required to assess the finding of individual tumor cells as an independent risk factor.

After resection of the primary tumor, which nowadays because of early diagnosis often measures less than 2 cm in diameter, the residual tumor cell mass can be assessed as minimal and may be less than 10^8 cells. The micrometastatic cells apparently represent a rather selected population of tumor cells resistant to the various stresses exerted on the cells during mobilization and transport through the circulation. Their quiescent state may be one of the reasons why the cells are resistant to so many chemotherapeutic agents. On the other hand, the absence of their own vascular system with the protective shield of the basal membrane could render the cells particularly vulnerable for immunologically targeted therapies. As a first approach to answer the question of accessibility of micrometastatic cells to infused monoclonal antibodies we tested whether micrometastatic cells could be labeled with monoclonal antibodies directed against 17-1A epithelial antigen. After infusion of 100–500 mg murine monoclonal antibody 17-1A, epithelial cells labeled with murine immunoglobulins were found in bone marrow. We therefore assume that in vivo labeling of disseminated tumor cells by macromolecular drugs such as antibodies or antibody conjugates may be feasible. In more than 20 patients longitudinal monitoring of epithelial cells in bone marrow yielded consistent results. In a few patients epithelial cells have now been detected over a period of up to 2 years.

With the unequivocal demonstration of viable tumor cells far from the site of primary tumor and in the absence of any clinical signs of manifest metastasis every effort is directed towards understanding the intrinsic and extrinsic factors governing the fate of these cells. Since many of the cancers described here are incurable after clinical metastasis has occurred, it may well be that the subclinical onset of metastasis is already a signum mortis for the patient. The early detection of these cells, however, gives hope that by identifying the enemy we may be better armed to fight it. At the same time we may spare patients from a potentially toxic treatment who are free of residual disease. Beyond its immediate clinical application the technique described here may shed light on a particularly dark stage of the malignant disease.

References

1. Boone T, Van Pel A, De Plaen E, Van den Eynde B, Hainant P (1989) Genes coding for TUM$^-$ transplantation antigens. A model for TSTA? In: Melchers F et al. (eds) Progress in Immunology, Vol 7. Springer, Berlin Heidelberg New York, pp 1063–1070
2. Moll R, Franke WW, Schiller DL, Geiger B, Krepler R (1982) The catalogue of human cytokeratins: pattern of expression in normal epithelia, tumors and cultured cells. Cell 31:11–24

3. Debus E, Weber K, Osborn M (1982) Monoclonal cytokeratin antibodies that distinguish simple from stratified squamous epithelia: characterization on human tissues. Embo J 1:1641–1647
4. Schlimok G, Funke I, Holzmann B, Göttlinger H, Schmidt G, Häuser H, Swierkot S, Warnecke HH, Schneider B, Koprowski H, Riethmüller G (1987) Micrometastatic cancer cells in bone marrow: in vitro detection with anti-cytokeratin and in vivo labelling with anti 17-1A-monoclonal antibodies. Proc Natl Acad Sci USA 84:8672–8676
5. Riethmüller G, Schlimok G, Lehmann JM, Stade BG, Funke I, Bock B, Schweiberer B, Johnson JP (1989) Immunological analysis of micrometastases and the metastatic phenotype of human tumors. In: Melchers F et al. (eds) Progress in Immunology, Vol. 7. Springer, Berlin Heidelberg New York, pp 1079–1086
6. Schlimok G, Funke I, Bock B, Schweiberer B, Witte J, Riethmüller G (1990) Epithelial tumor cells in bone marrow of patients with colorectal cancer: immunocytochemical detection, phenotypic characterization and prognostic significance. J Clin Oncol 8: 831–837
7. Schlimok G, Funke I, Strobel F, Witte J, Riethmüller G (1990) Comparative analysis of monoclonal antibodies for the detection of epithelial micrometastatic tumor cells in bone marrow and its clinical application to gastric cancer. (in press)
8. Weiss L, Grundmann E, Torhorst J et al. (1986) Haematogenous metastatic patterns in colonic carcinoma: an analysis of 1541 necropsies. J Pathol 150:195–203
9. Funke I, Schlimok G, Bock B, Schweiberer B, Riethmüller G (1989) Micrometastasis: characterization and tumorigenicity of disseminated tumor cells in human bone marrow. In: Osborne M, Weber K (eds.) Current communications in molecular biology. Cold Spring Harbor Laboratory, New York, pp 214–244
10. Delsol G, Gatter KC, Stein H, Erber WN, Pulford KAF, Zinne K, Mason DY (1984) Human lymphoid cells express epithelial membrane antigen. Lancet ii:1124–1129

The Microvasculature of Free Pancreatic Islet Grafts

M.D. Menger, F. Hammersen†, and K. Messmer

Introduction

Although diabetes mellitus is one of the principal civilization diseases, there is still no curative treatment. The enthusiasm for approaching this goal by the development of whole-pancreas transplantation as a clinical routine [1–3] is restricted both by the small number of donor organs available compared to the urgent need of organs due to the large number of diabetic patients, and by the first posttransplant long-term experiences with limited effect on the typical late complications of diabetes, such as nephropathy, retinopathy, and neuropathy [4–6]. Moreover, recently the justification of pancreas transplantation for use in the treatment of diabetes mellitus has been questioned in general [7].

The idea of transplanting pancreatic islets of Langerhans devoid of exocrine tissue as a free graft has attracted attention since Moskalewski [8] described an enzymatic method for isolating pancreatic islets by means of collagenase digestion. The first experience with experimental islet transplantation was reported in 1970, when Younoszai and coworkers [9] ameliorated chemically induced diabetes in the rat by intraperitoneal transplantation of isolated islets. In the same year, Reemtsma [10] transplanted fish islets intramuscularly, into the eye, and contained within millipore chambers into the peritoneal cavity of diabetic rats. The first major exercise in experimental islet transplantation was reported in 1972 [11] and resulted in long-term amelioration of diabetic phenomena such as hyperglycemia, polyuria, and glucosuria after intraperitoneal and intramuscular isotransplantation of 400–600 islets into streptozotocin-diabetic rats.

During the following years, a large number of successful experimental studies on islet transplantation were performed, particularly after implantation in highly vascularized organs such as the liver [12], spleen [13], and the kidney subcapsular tissue [14]. The optimal site for islet transplantation has still to be defined [15]. The less vascularized sites, such as subcutaneous and intramuscular tissue, are regarded as less suited for long-term implantation of islet tissue [16], and the intraperitoneal site has the disadvantage of requiring large numbers of isolated islets to obtain reversal of hyperglycemia [11,12]. Using highly vascularized sites of implantation a number of studies have compared portal versus systemic venous drainage of islet grafts and suggest that less islet

K. Messmer M. Stein (Eds.)
Pathways in Applied Immunology

tissue is needed to reverse diabetes and improve metabolic control if the islets are draining into the portal rather than the systemic venous system [17–19]. Since clinical islet transplantation at present involves marginal quantities of functioning islets, the site with portal venous drainage is considered as the most appropriate choice [15].

In addition, the ideal transplantation site has been studied experimentally with regards to immunological privilege, and areas such as the cerebral ventricle [20] and testis [21–23] have been shown to be superior, while the kidney subcapsular space was suggested as conveying some immunological advantages as compared to the liver or the spleen [14,24].

Experimental islet transplantation has been performed frequently with sufficient functional results also in allogeneic systems [25–34] as well as in xenogeneic systems [22,35–38] using different techniques to prevent graft rejection; however, the initial clinical trials proved unsuccessful [39–42]. Although there might be considerable excitement at present due to preliminary reports on new experiences with clinical islet transplantation [43–45], a functional success rate like in whole-organ grafts such as the kidney, the heart, and the liver has not yet been obtained in free pancreatic islet grafts.

In contrast to whole-organ grafts, which are revascularized at the time of transplantation, freely grafted islets require the process of angiogenesis and revascularization to establish the microcirculatory network necessary for sufficient nutritional blood supply. Because of the lack of adequate models for experimental studies, there is only little information available concerning the microvasculature of free pancreatic islet grafts [46,47]. Therefore, we present in detail features of the process of revascularization of free pancreatic islet grafts, using a novel technique, which allows in vivo observation of the islets' microcirculation over a prolonged period of time [48].

Methods

Skinfold Preparation. For our studies we used the hamster dorsal skinfold chamber (Fig. 1a), which allows observation of the microcirculation in the awake animal over a long period of time by means of intravital fluorescence microscopy. The tissue within the chamber consists of one layer of striated muscle and one layer of skin. The chamber technique and its implantation procedure have been described previously in detail [49]. Briefly, under Nembutal anesthesia (50 mg/kg body wt i.p.) two symmetrical titanium frames are implanted on the extended dorsal skinfold of 8- to 10-week-old Syrian golden hamsters (60–80 g) so that they sandwich the double layer of skin. One layer is then completely removed in a circular area of 15 mm in diameter. The remaining underlying tissue, consisting of a thin layer of skin muscle (cutaneus maximus muscle) and subcutaneous tissue, is covered with a removable cover slip, incorporated in one of the titanium frames. After a recovery time of 48 h to exclude pathological effects on the microcirculation due to anesthesia and/or surgical trauma, intravital fluorescence microscopy can be performed

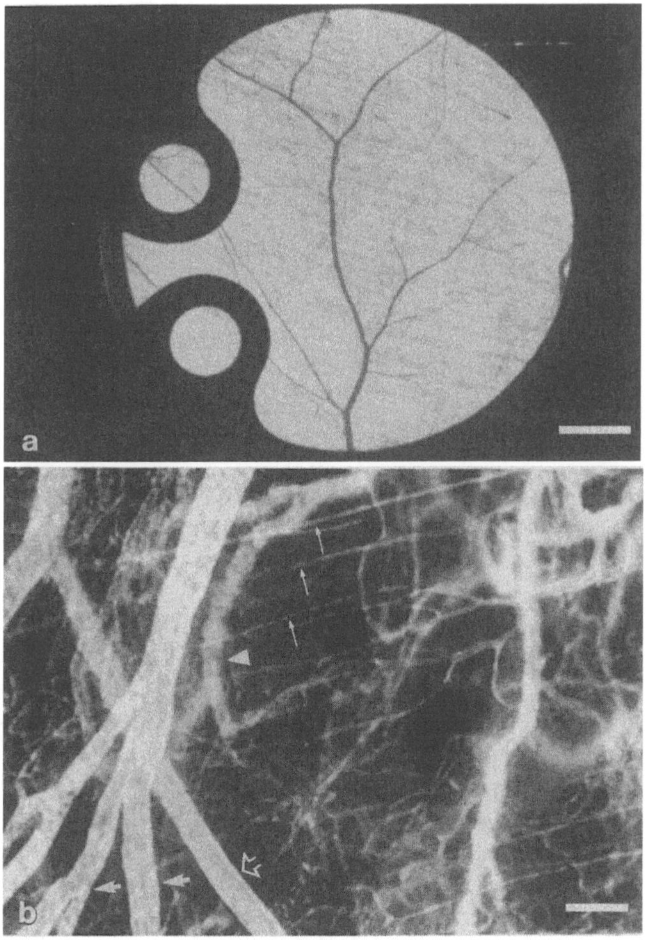

Fig. 1. **a** Hamster dorsal skinfold chamber. The tissue within the chamber consists of one layer of skin (striated) muscle and subcutis and allows intravital fluorescence microscopy in the awake animal over a prolonged period of time; *bar*, 2 mm. **b** Intravital fluorescence microscopy of the hamster dorsal skinfold chamber demonstrating the microvascular segments of the preparation, i.e., arterioles (⇨), capillaries (→), postcapillary venules (➛) and collecting venules (➝). Contrast enhancement is achieved by intravenous application of 5% FITC-dextran 150 000; *bar*, 100 μm

after intravenous injection of 5% FITC-dextran 150 000 (Sigma, St. Louis MO, USA) over a period of up to 4 weeks (Fig. 1b).

Islet Isolation and Islet Transplantation. Pancreatic islets were isolated from Syrian golden hamsters for syngeneic transplantation and from DA rats (dark Agouti; RTl[a]) for xenogeneic transplantation using vital staining with neutral red and a modified collagenase digestion technique [50]. This technique allows isolation of 500–800 islets from one donor pancreas. A few islets of each

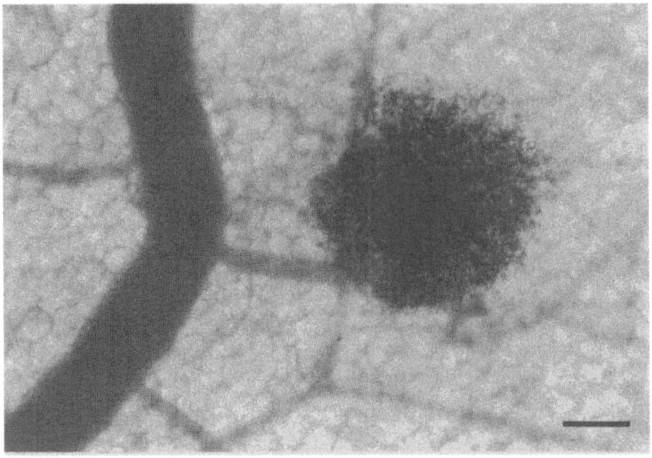

Fig. 2. Isolated pancreatic islet 1 h after transplantation into the hamster dorsal skinfold chamber; in vivo staining with neutral red; *bar*, 50 μm

sample were used to verify their viability by means of thiazole tetrazolium [51] and dithizon [52] staining. A hand-picking procedure guaranteed single exocrine free islets for transplantation. For islet transplantation, the cover glass of the skinfold chamber was removed and 10–15 islets were placed on the striated muscle tissue within the chamber (Fig. 2). This technique allows repetitive intravital microscopic observations of the microvasculature of each single pancreatic islet graft.

Histological Examinations. The dorsal skinfold preparation was fixed by both vascular perfusion and superfusion with 2.5% glutaraldehyde in 0.1 *M* sodium cacodylate buffer (pH 7.4) for 30 min. After another fixation for 24 h in the same fixative, small pieces, each containing one islet, were dissected under a stereomicroscope. Electron microscopy was performed, employing a Zeiss-EM 10A microscope operated at 60 kV. Semithin sections (0.5–1 μm) stained with methylene blue–azur II were used for light microscopy. The presence of intracellular insulin was demonstrated immunohistochemically by means of an anti-insulin staining technique [53].

Revascularization of Syngeneic Grafts

Angiogenesis of Syngeneic Islet Grafts. Between days 2 and 4 after transplantation onto striated muscle tissue, more than 90% of syngeneic islet grafts demonstrate the first signs of revascularization. Angiogenesis in its early phase is characterized by the protrusion of capillary sprouts, originating from the venular segments of the skin muscle capillaries and postcapillary venules (Fig. 3). This capillary sprouts, frequently in the configuration of sinusoidal

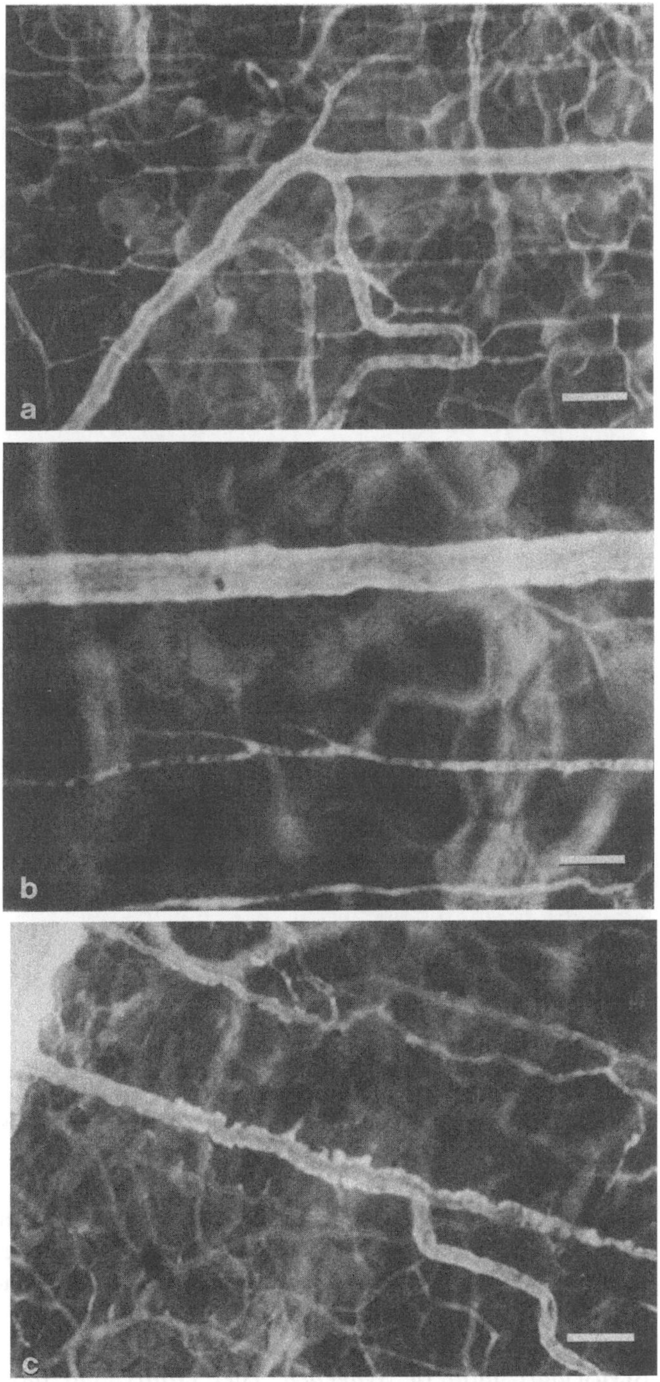

Fig. 3. a Striated muscle capillaries (running parallel) and postcapillary venules of the hamster dorsal skinfold chamber. **b** Higher magnification reveals the normal intact microvascular wall of the capillaries and postcapillary venules. **c** Capillaries and postcapillary venules of the striated muscle layer of the hamster dorsal skinfold chamber within an area where a syngeneic pancreatic islet was transplanted 2 days before. **Figure 3d,e** see p. 114

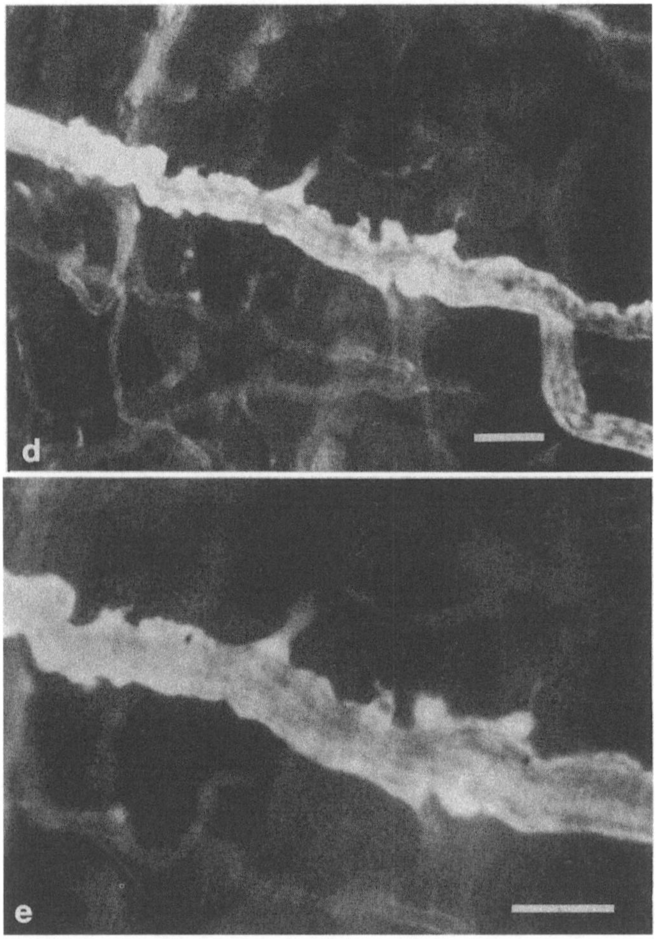

Fig. 3. d, e Higher magnification reveals the very beginning of angiogenesis and revascularization, i.e., capillary sprouts, demonstrated by irregularities of the microvessel wall of the capillaries and postcapillary venules of the host tissue. Intravital fluorescence microscopy, contrast enhancement by 5% FITC-dextran 150000 i.v.; *bar*, 100 μm and 50 μm, respectively

structures, developing from multiple host vessels within the area of the transplanted islet (Fig. 4a). Over the following days the growth of these capillary sprouts can be observed; the sprouts connect with each other and with additional microvessels of the host tissue (Fig. 4b).

Revascularization of Syngeneic Islet Grafts. On day 6 after transplantation a growing microvascular (capillary) network can be observed revealing the morphological micro-angio-architecture characteristic of pancreatic islets (Fig. 4c). Over the following days perfusion of these newly formed vessels increases, as does the density of the whole microvascular network. On days 10–14 the process of revascularization of syngeneic islet grafts is completed;

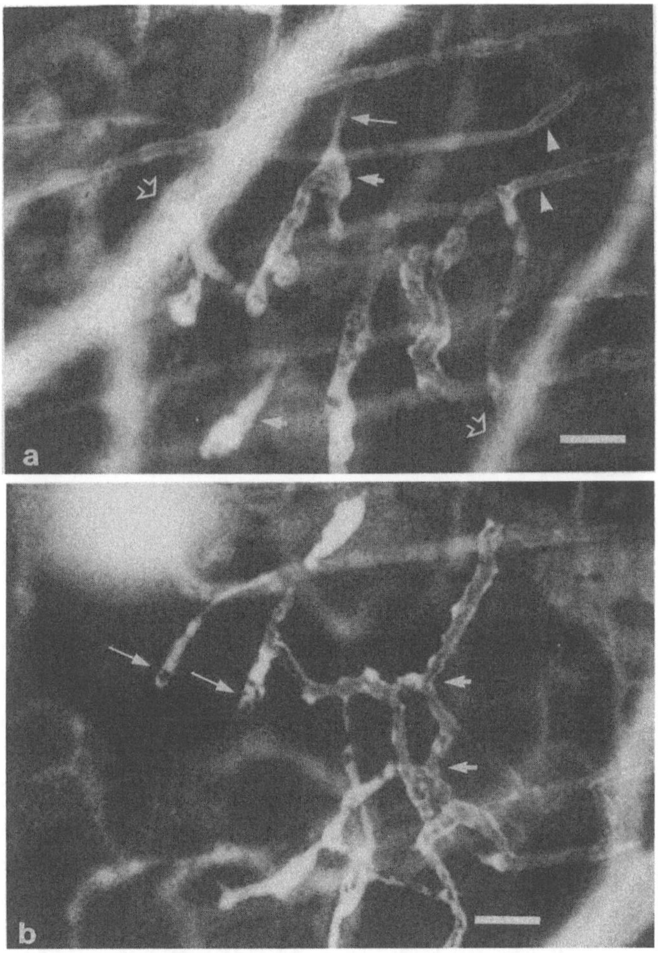

Fig. 4. **a** Syngeneic pancreatic islet 3 days after transplantation into the hamster dorsal skinfold chamber. The process of revascularization is characterized by sinusoidal microvascular formations (➧) and capillary sprouts (→) originating from the venular segments of the striated muscle capillaries (▶) and the postcapillary venules (▷) of the host's microvasculature. **b** On day 4 intercapillary anastomoses can be observed (➧), while further sprouting of new capillaries (→) demonstrates the persistent growth of the microvascular network

a glomeruluslike tuft of a microvascular network is apparent, very similar to the morphology of the microvasculature of pancreatic islets in situ (Fig. 5a). Intravital microscopic analyses using in vivo staining of white blood cells with acridine orange revealed the absence of inflammatory reactions in syngeneic islet grafts, as indicated by normal flow behavior of the white blood cells [54].

Structural Analysis of the Newly Formed Microvascular Network of Syngeneic Islet Grafts. The structural analysis of the microvascular network of the pancreatic islet grafts regularly reveals one or two arterial supply channels, con-

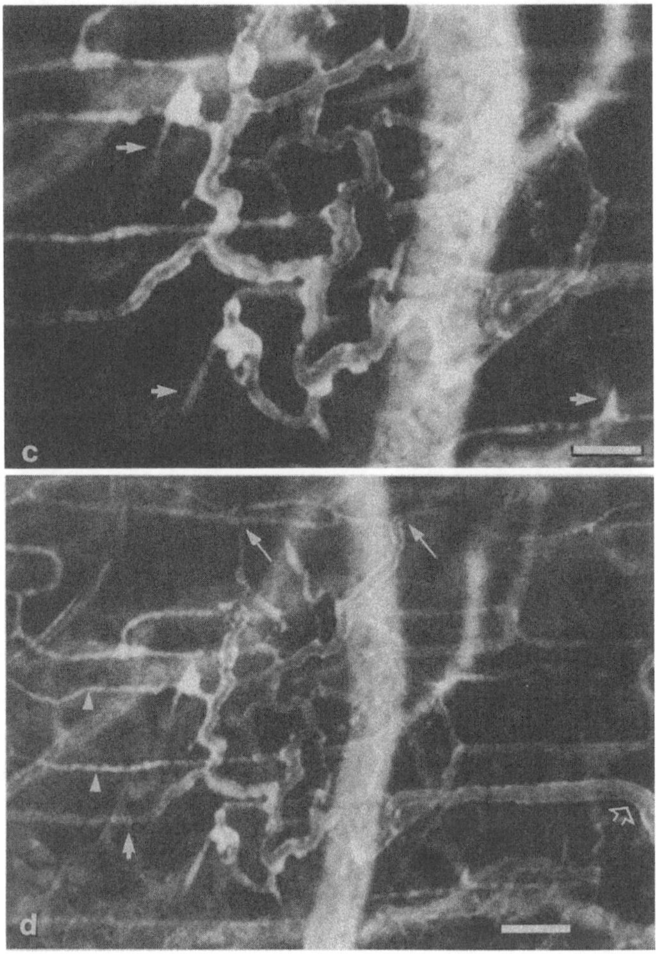

Fig. 4. c On day 6 after transplantation the newly formed microvessels create a network of capillaries with low density. Further capillary sprouts indicate that the process of revascularization is still not terminated (➧). **d** The overview of the identical islet on day 6 shows the interaction of the newly formed microvessels with the microvasculature of the host tissue, demonstrating an arteriole, serving as vascular supply (➧, former striated muscle capillary, running parallel with the other striated muscle capillaries ▶), the glomeruluslike capillary network, and a postcapillary venule (⇨). In addition, intercapillary anastomoses between the islet and striated muscle capillaries can be observed (→). Intravital fluorescence microscopy, contrast enhancement by 5% FITC-dextran 150000 i.v.; *bar*, 50 μm and 100 μm, respectively

sisting of former capillaries of the skin muscle, but with altered hemodynamic conditions (Fig. 5b). The diameters of these vessels increase up to three fold, and red blood cell velocity can be measured within the range of 0.6–1.5 mm/s, while such velocities in normal capillaries of the skin muscle range from 0.2 to 0.3 mm/s [55]. The glomeruluslike tuft of islet capillaries can be differen-

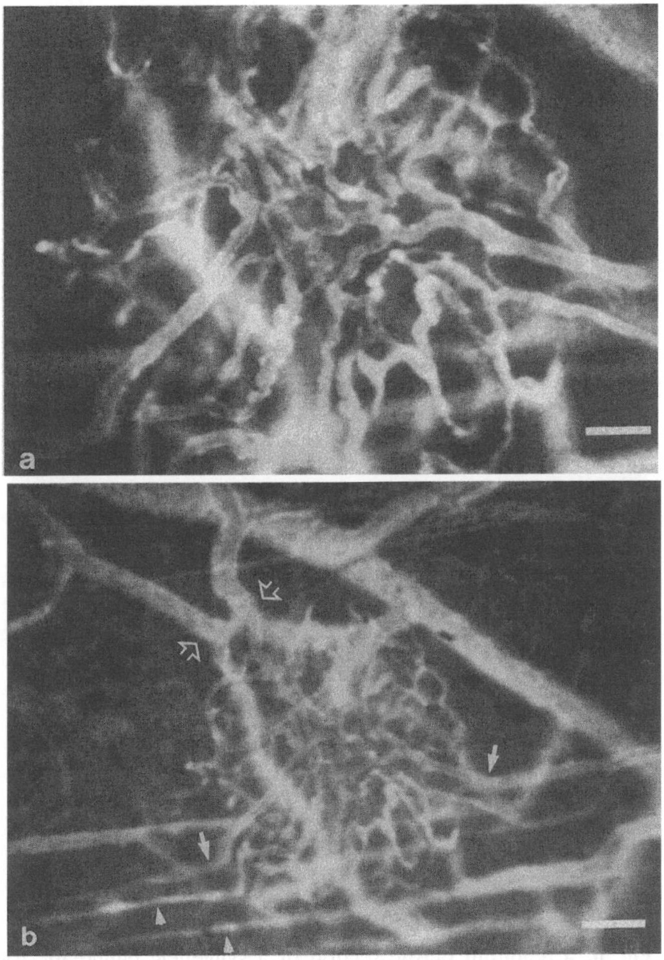

Fig. 5a,b. Syngeneic pancreatic islet graft 10 days after transplantation into a hamster dorsal skinfold chamber. **a** The microvasculature of the islet graft consists of a dense glomeruluslike network of capillaries. **b** The overview of the islet demonstrates the location of the islets' microvasculature within the microvasculature of the host tissue. The islet is supplied by two arterioles (➧, former striated muscle capillaries, running parallel with the other striated muscle capillaries ▶), and drained by postcapillary venules (⇨). Intravital fluorescence microscopy, contrast enhancement by 5% FITC-dextran 150 000 i.v.; *bar*, 50 μm and 100 μm, respectively

tiated into afferent and efferent capillary loops, and consistently a venular draining system can be observed (Fig. 5b). However, blood from the islet grafts is not exclusively drained into postcapillary venules, but, in analogy to the insuloacinar portal system, intercapillary anastomoses between islet and host tissue (muscle) capillaries can be observed [56]. Figure 5b gives an overview of the microvasculature of an islet graft located on the surface of the skin muscle (host tissue) with its entire microvascular supply.

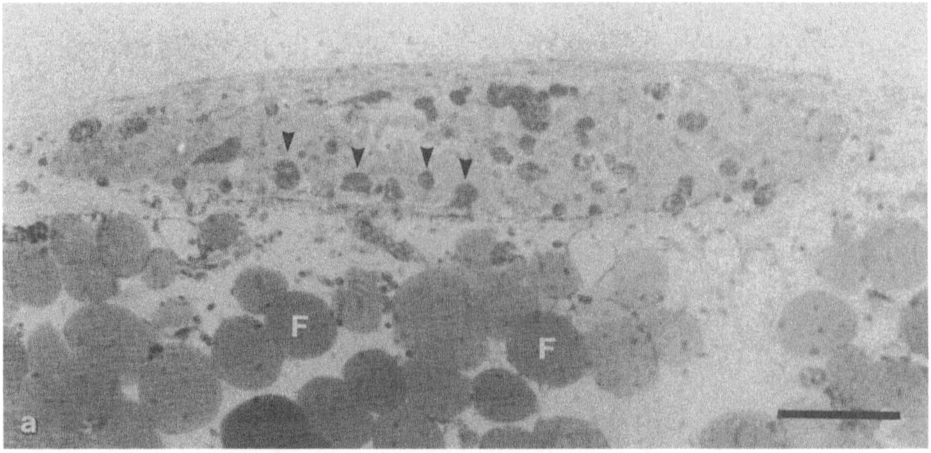

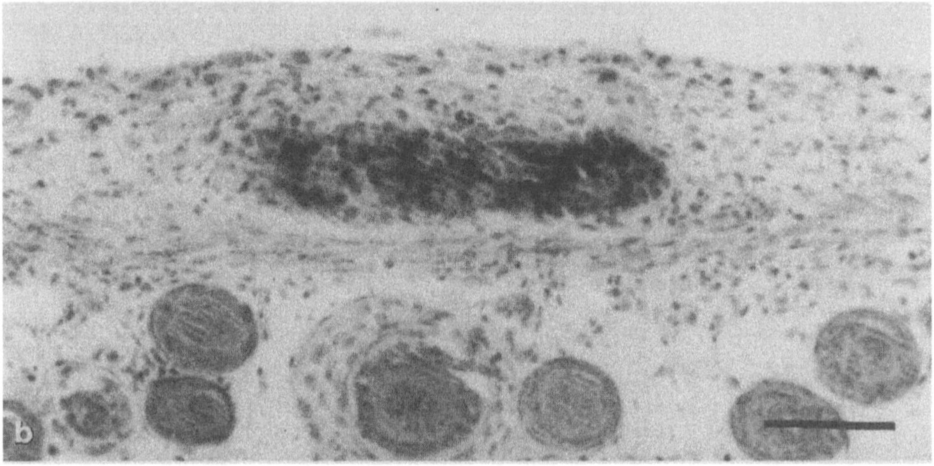

Fig. 6. a Semithin plastic section of a syngeneic pancreatic islet graft 10 days after transplantation into a hamster dorsal skinfold chamber. Note the density of microvessels within the islet graft (▶). *F*, Fat cells. Methylene blue–azur II staining; *bar*, 50 µm. **b** Paraffin section of a syngeneic islet graft 10 days after transplantation. The anti-insulin immunocytochemical technique reveals nearly homogeneous distribution of insulin in most of the darker staining islet cells; *bar*, 100 µm. (From [48] with permission)

Fig. 7. a Syngeneic islet 3 days after grafting into a hamster dorsal skinfold chamber. The islet cells show entirely normal fine structure (⋆) with numerous mitochondria together with cisternae of the RER, and some of the cells (*) display already their characteristic secretory granules (see also below). Notice complete absence of blood vessels within the graft (×2800). **b** Higher magnification (×9700) of the central part of the graft from above clearly shows that all the bizarrly deformed RBCs (*E*) are tightly wedged between the islet cells from which they are not separated by a basal lamina. Notice normal fine structure of organelles and secretory granules (*) of the islet cells

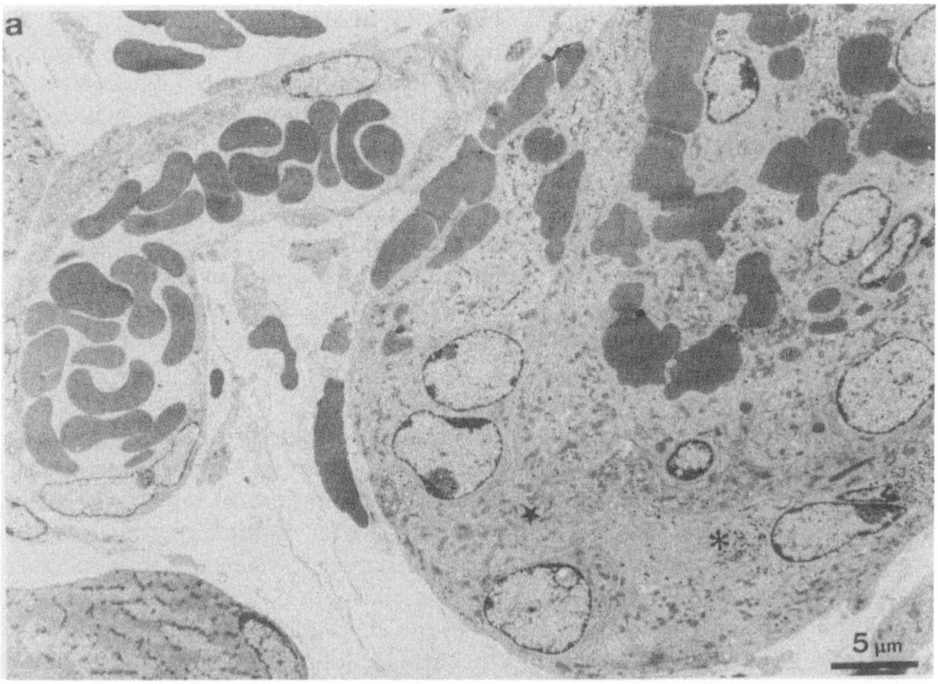
a
5 µm

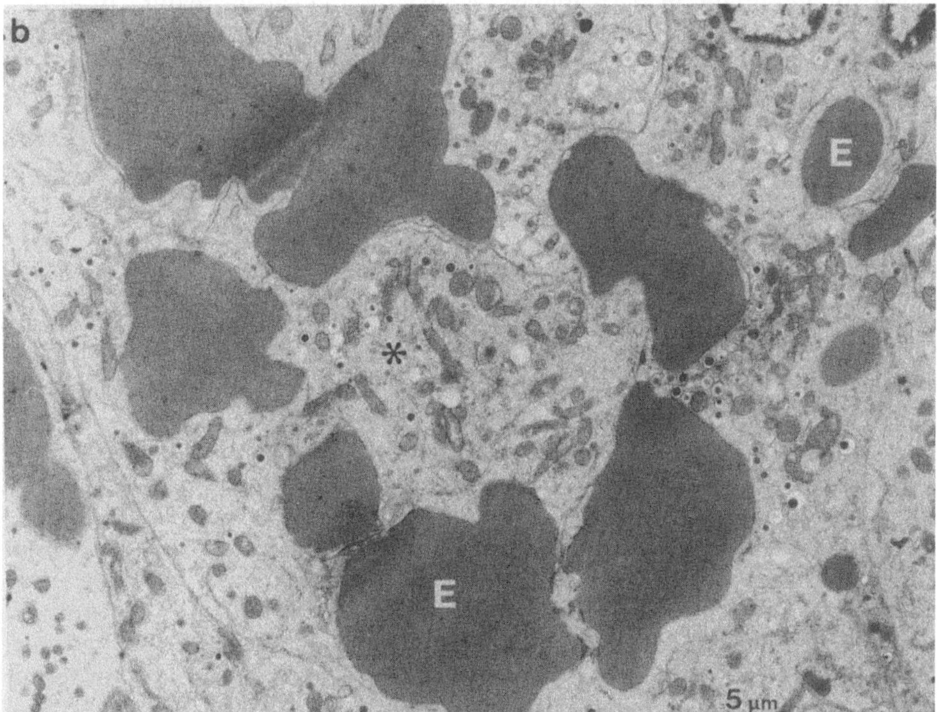
b
E
E
5 µm

Histology of Syngeneic Islet Grafts. Histology of syngeneic islet grafts reveals a regular structure of the cellular components in conjunction with densely arranged microvessels (Fig. 6a). Intracellular insulin content was confirmed on days 10, 14, 18, and 24 after transplantation and showed homogeneous distribution of intracellular insulin (Fig. 6b) The regular morphological structure of the syngeneic islet grafts was also confirmed by electron microscopy [48].

In addition, this study has revealed that the mechanisms of revascularization are not based on inosculation (anastomosis of graft vessels with host vessels) and penetration of endothelial cells into the preexisting graft microvessels, as suggested by others [57,58] who have studied revascularization of skin grafts, but ingrowth of newly formed microvessels into the islet grafts is encountered (Fig. 7).

Angiogenesis and Revascularization of Free Islet Xenografts. The time course of revascularization of xenogeneic islet grafts is very similar to that of syngeneic grafts during the first 6 days after transplantation. However, when the microcirculatory network is established, i.e., after 6–10 days, host-versus-graft reaction can be observed, demonstrating microvascular phenomena such as deterioration of microvascular blood flow, formation of perivascular edema, and accumulation of white blood cells within the microvasculature of the islet xenografts. It should be emphasized that white blood cell – endothelium interaction, preferentially in postcapillary venules, is the salient feature from day 10 after transplantation, followed by emigration of these cells through the endothelial wall [59]. The consequence of these sequelae is the breakdown of the islet microcirculation with, finally, cessation of blood flow and destruction of the microvessels, which usually starts in the central area of the xenograft islet's capillary network (Fig. 8).

Discussion

Transplantation of insulin-secreting tissue as a free graft has the potential to be a safe and simple procedure for curing diabetes. However, while in numerous experimental studies on transplantation of isolated pancreatic islets diabetes was successfully reversed, the clinical attempts to achieve insulin independence have failed. The major difference between vascularized whole-organ grafts – such as those of heart, liver, or kidney, which are successfully transplanted in man – and free islet grafts is the lack of nutritive capillaries within the islets, which therefore depend upon diffusion for their nutritional supply for the first few days after transplantation and later upon the time course and quality of the process of neovascularization.

Although nutritional blood supply is of crucial importance for a transplant, little is known about the microvasculature of free islet grafts. In recent years only few studies on microvascular blood flow of free pancreatic islet grafts have addressed the process of angiogenesis and revascularization [60,48] and conditions of islets or external factors such as hyperglycemia [46,61],

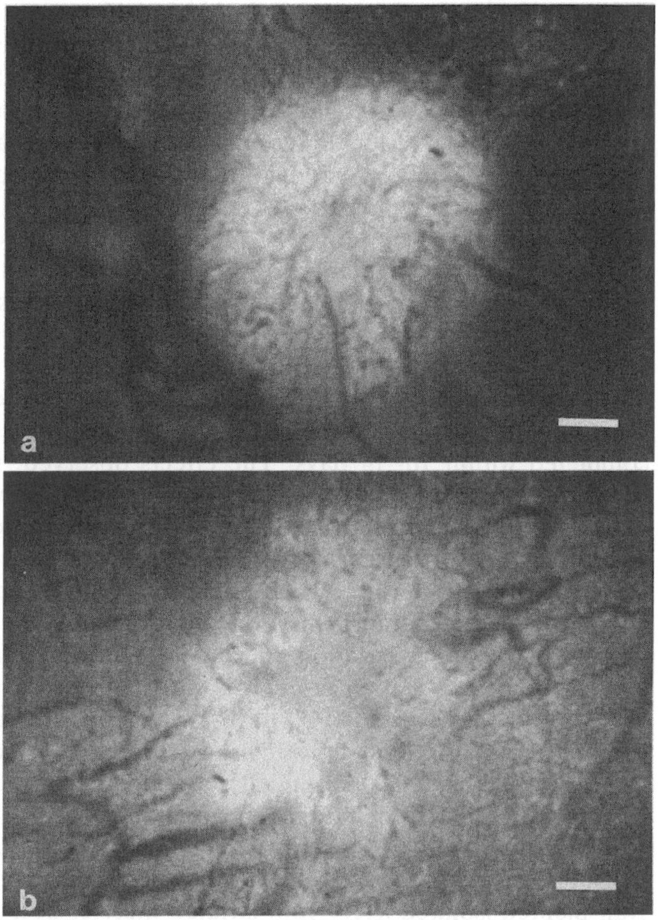

Fig. 8. a Glomeruluslike network of a syngeneic pancreatic islet graft 10 days after transplantation. Using the in vivo staining technique with the low molecular weight fluorescent marker acridine orange, the microvessels appear dark, while the islet tissue is stained brightly. **b** A xenogeneic pancreatic islet graft 10 days after transplantation using the same staining technique. Note the loss of microvessels, particularly within the central area of the islet, indicating the process of rejection. Intravital fluorescence microscopy, contrast enhancement by 1% acridine orange i.v.; *bar*, 100 μm

culture [62], and immunosuppression [47,59,63,64] on the islets microvascular blood flow. There seems to be agreement that the process of revascularization of free islet grafts is completed after 10–14 days, which is in accordance with the findings from histological studies [65]. Hyperglycemia does not alter angiogenesis of free islet grafts, as we have demonstrated by intravital microscopy [61]. However, contrary results were reported by Sandler and Jansson [46]; these authors used the microsphere technique, an indirect method for measuring nutritional flow, precluding continuous and repetitive measurements. Furthermore, their measurements of islet graft microvascular blood flow by means of microspheres were performed 1 months after transplantation, whereas angiogenesis and revascularization take place during the first 10–14 days after transplantation. The conclusion of Sandler and Jansson [46] that microvascular blood flow of islet grafts is reduced due to hyperglycemia does not allow one to infer effects of hyperglycemia on angiogenesis and revascularization.

Apart from our own studies, there exists only one report on the microvasculature of free syngeneic islet grafts based on intravital microscopy. Rooth et al. [47] analyzed the effect of prednisolone and cyclosporin A on vascular ingrowth after transplantation of free islet clusters under the kidney capsule. While prednisolone treatment did not influence revascularization of the islet clusters, a detrimental effect on vascular ingrowth was observed with cyclosporin A, which, however, could be prevented with verapamil. It should be noted that analyses of microvascular blood flow were performed not earlier than 1 month after transplantation, while the process of revascularization is completed within 10–14 days. The effects observed, in particular with cyclosporin A, might not be due to suppression of angiogenesis but to a toxic effect on the microvascular endothelium after revascularization has taken place. This would be in accordance with in vitro observations indicating endothelial cell injury upon cyclosporine, which might also be a mechanism favoring vascular complications after transplantation [66]. The effects of cyclosporin A on the process of neovascularization and on the microvascular endothelium are presently under investigation in our laboratory.

Rooth and coworkers [47] have also studied the influence of prednisolone on microvascular blood flow in islet clusters transplanted under the kidney capsule, and they have excluded negative influences of corticoid treatment on the islet microcirculation. The authors concluded that revascularization (capillary ingrowth) of free pancreatic islet grafts is not hampered by prednisolone treatment. However, these results could not be confirmed when we transplanted isolated islets onto striated muscle, because in these experiments angiogenesis and revascularization were found delayed under treatment with 5 mg/kg prednisolone [67]. The different results obtained in these two studies may be due to differences in design of the protocols. While we have analyzed angiogenesis of islet grafts during the first 14 days after transplantation, i.e., the time period of revascularization, Rooth et al. [47] performed their analyses of microvascular blood flow 1 month after transplantation. In addition, the model used by Rooth and coworkers [47] has disadvantages, because it does not allow sequential measurements and is confined to the analysis of islet clusters, potentially including microvessels of the granulation tissue under the kidney capsule. In contrast, after islet transplantation in our observation window, the microvasculature of each single islet graft can be analyzed, also under treatment with prednisolone [48,67].

Xenogeneic islet grafts reveal a similar process of revascularization during the first 6 days after transplantation, however, as soon as the microvasculature is restored, rejection phenomena are observed [59]. Accumulation of white blood cells and their interaction with the microvascular endothelium appear to play a major role in the host-versus-graft reaction. Neither (±)-15-

deoxyspergualin nor cyclosporin A were able to substantially influence the rejection phenomena or to improve the process of revascularization [63,64]. Further studies on revascularization of islet grafts are necessary to identify new therapeutic directives to improve revascularization and to prevent graft rejection. These strategies should include anti-lymphocyte serum, as well as monoclonal antibodies directed against the endothelium/leukocyte adhesion molecules and the intracellular adhesion molecules, which are considered to be involved in leukocyte endothelium interaction and leukocyte emigration to the host tissue [68].

References

1. Sutherland DER, Moudry KC (1987) Pancreas transplant registry report. Transplant Proc 19 [Suppl 4]:5–7
2. Tydén G, Brattström C, Bolinder J, Reinholt F, Östman J, Lundgren G, Wilczek H, Groth CG (1988) Pancreatic transplantation in diabetics with preuremic nephropathy. Transplant Proc 20:471–474
3. Sutherland DER, Kendall DM, Moudry KC, Navarro X, Kennedy WR, Ramsay RC, Steffes MW, Mauer SM, Goetz FC, Dunn DL, Najarian JS (1988) Pancreas transplantation in nonuremic, type I diabetic recipients. Surgery 104:453–464
4. Ramsay RC, Goetz FC, Sutherland DER, Mauer SM, Robison LL, Cantrill HL, Knobloch WH, Najarian JS (1988) Progression of diabetic retinopathy after pancreas transplantation for insulin-dependent diabetes mellitus. N Engl J Med 318:208–214
5. Solders G, Wilczek H, Gunnarson R, Tydén G, Persson A, Groth C-G (1987) Effects of combined pancreatic and renal transplantation on diabetic neuropathy; a two-year follow-up study. Lancet ii:1232–1235
6. Groth C-G (1988) Pancreas transplantations. Lancet i:1400 (letter)
7. Pyke D (1988) Pancreas transplantation for diabetes? Lancet i:816–817
8. Moskalewsky S (1965) Isolation and culture of islets of Langerhans of the guinea pig. Gen Comp Endocrinol 5:342–343
9. Younoszai R, Sorenson RL, Lindall AW (1970) Homotransplantation of isolated pancreatic islets. Diabetes 19 [Suppl 1]:406
10. Reemtsma K (1970) Experimental islet cell grafting: a transplantation model. Transplant Proc 2:513–515
11. Ballinger WF, Lacy PE (1972) Transplantation of intact pancreatic islets in rat. Surgery 72:175–186
12. Kemp CB, Knight MJ, Scharp DW, Ballinger WF, Lacy PE (1973) Effect of transplantation site on the results of pancreatic islet isografts in diabetic rats. Diabetologia 9:486–491
13. Finch DR, Wise PH, Morris PJ (1977) Successful intrasplenic transplantation of syngeneic and allogeneic isolated pancreatic islets. Diabetologia 13:195–199
14. Reece-Smith H, DuToit DF, McShane P, Morris PJ (1981) Prolonged survival of pancreatic islet allografts transplanted beneath the renal capsule. Transplantation 31: 305–306
15. Gray DWR, Morris PJ (1987) Developments in isolated pancreatic islet transplantation. Transplantation 43:321–331
16. Gray BN, Watkins E (1976) Isolated islet transplantation in experimental diabetes. Aust J Exp Biol Med Sci 54:57–61
17. Brown J, Mullen Y, Molnar IG, Clark WR (1976) Importance of hepatic-portal route for control of diabetes. Diabetes 25 [Suppl 1]:338 (abstr)

18. Matas AJ, Payne WD, Grotting JC, Sutherland DER, Steffes MW, Hertel BF, Najarian JS (1977) Portal versus systemic transplantation of dispersed neonatal pancreas. Transplantation 24:333–337
19. Squifflet JP, Sutherland DER, Florack G, Morrow CE, Najarian JS (1985) Physiologic comparison of segmental pancreas and islet transplants in rats. Transplant Proc 17: 378–380
20. Tze WJ, Tai J (1986) Intrathecal allotransplantation of pancreatic endocrine cells in diabetic rats. Transplantation 41:531–534
21. Ferguson J, Scothorne J (1977) Extended survival of pancreatic islet allografts in the testis of guinea-pigs. J Anat 124:1–8
22. Bobzien B, Yasunami Y, Majercik M, Lacy PE, Davie JM (1983) Intratesticular transplants of islet xenografts (rat to mouse). Diabetes 32:213–216
23. Selawry HP, Whittington K (1984) Extended allograft survival of islets grafted into intra-abdominally placed testis. Diabetes 33:405–406
24. Toledo-Pereyra LH, Bandlien KO, Gordon DA, Mackenzie GH, Reyman TA (1984) Renal subcapsular islet cell transplantation. Diabetes 33:910–914
25. Bowen KM, Andrus L, Lafferty K (1980) Successful allotransplantation of mouse pancreatic islets to non immunosuppressed recipients. Diabetes 29 [Suppl 1]:98–104
26. Lacy PE, Davie JM, Finke EH (1979) Prolongation of islet allograft survival following in vitro culture (24 degrees C) and a single injection of ALS. Science 204:312–313
27. Faustman DL, Steinman RM, Gebel HM, Hauptfeld V, Davie JM, Lacy PE (1984) Prevention of rejection of murine islet allografts by pretreatment with anti-dentritic cell antibody. Proc Natl Acad Sci USA 81:3864–3868
28. Lau H, Reemtsma K, Hardy MA (1984) Prolongation of rat islet allograft survival by direct ultraviolet irradiation of the graft. Science 223:607–609
29. Hardy MA, Lau H, Weber C, Reemtsma K (1984) Pancreatic islet transplantation. Induction of graft acceptance by ultraviolet irradiation of donor tissue. Ann Surg 200: 441–448
30. Lau H, Reemtsma K, Hardy MA (1984) The use of direct ultraviolet irradiation and cyclosporine in facilitating indefinite pancreatic islet acceptance. Transplantation 38: 566–569
31. Faustman D, Hauptfeld V, Lacy P, Davie J (1981) Prolongation of murine islet allograft survival by pretreatment of islets with antibody directed to Ia determinants. Proc Natl Acad Sci USA 78:5156–5159
32. Altman JJ, Houlber D, Callard P, McMillan P, Solomon BA, Rosen J, Galletti PM (1986) Long-term plasma glucose normalization in experimental diabetic rats with macroencapsulated implants of benign human insulinomas. Diabetes 35:625–633
33. Sun AM, O'Shea GM, Goosen MF (1984) Injectable microencapsulated islet cells as a bioartificial pancreas. Appl Biochem Biotechnol 10:87
34. Taunton-Rigby A (1986) Transplantation of encapsulated islets. Transplant Proc 18: 1841–1844
35. Lacy PE, Davie JM, Finke EH (1980) Prolongation of islet xenograft survival without continuous immunosuppression. Science 209:283–285
36. Yasunami Y, Lacy PE, Davie JM, Finke EH (1983) Use of in vitro culture at 37 degrees centigrade to prolong islet xenograft survival (rat to mouse). Transplant Proc 15:1371–1372
37. Tze WJ, Tai J (1983) Prolongation of islet allograft and xenograft in nonimmunosuppressed rat recipients. Metabolism 32:279–282
38. Yasunami Y, Lacy PE, Davie JM, Finke EH (1983) Prolongation of islet xenograft survival (rat to mouse) by in vitro culture at 37°C. Transplantation 35:281–284
39. Najarian JS, Sutherland DER, Matas AJ, Steffes MW, Simmons RL, Goetz FC (1977) Human islet transplantation: a preliminary report. Transplant Proc 9:233–236
40. Sutherland DER, Matas AJ, Goetz FC, Najarian JS (1980) Transplantation of dispersed pancreatic islet tissue in humans: autografts and allografts. Diabetes 29 [Suppl 1]:31–44
41. Kolb E, Largiader F (1980) Clinical islet transplantation. Transplant Proc 12 [Suppl 2]:205–207

42. Sutherland DER, Kendall D (1985) Clinical pancreas and islet transplant registry report. Transplant Proc 17:307–311
43. Farkas G, Karacsonyi S, Szabo M, Kaiser G (1987) Results of cultured fetal pancreatic islet transplantation in juvenile diabetic patients. Transplant Proc 19:2352–2353
44. Shumakov VI, Bljumkin VN, Ignatenko N, Skaletsky NN, Slovesna TA, Babikova RA (1987) The principal results of pancreatic islet cell culture transplantation in diabetes mellitus patients. Transplant Proc 19:2372
45. Warnock GL, Kneteman NM, Ryan EA, Evans MG, Seelis REA, Halloran PF, Rabinovitch A, Rajotte RV (1989) Continued function of pancreatic islets after transplantation in type I diabetes. Lancet ii:570–572
46. Sandler S, Jansson L (1987) Blood flow measurements in autotransplanted pancreatic islets of the rat. Impairment of the blood perfusion of the graft during hyperglycemia. J Clin Invest 80:17–21
47. Rooth P, Dawidson I, Lafferty K, Diller K, Armstrong J, Pratt P, Simonsen R, Täljedal I-B (1989) Prevention of detrimental effect of cyclosporin A on vascular ingrowth of transplanted pancreatic islets with verapamil. Diabetes 38 [Suppl 1]:202–205
48. Menger MD, Jäger S, Walter P, Hammersen F, Messmer K (1990) A novel technique for studies on the microvasculature of transplanted islets of Langerhans in vivo. Int J Microcirc Clin Exp 9:103–117
49. Endrich B, Asaishi K, Götz A, Messmer K (1980) Technical report – a new chamber technique for microvascular studies in unanesthetized hamsters. Res Exp Med 177: 125–134
50. Lacy PE, Kostianovsky M (1967) Method for the isolation of intact islets of Langerhans from the rat pancreas. Diabetes 16:35–39
51. Kühn F, Lippert H, Gütte N, Wolff H (1988) An in vitro rapid colorimetric assay for viability control of isolated pancreatic human islets before transplantation. Langenbecks Arch Chir Forum 88:97–99
52. Latif ZA, Noel J, Alejandro R (1988) A simple method of staining fresh and cultured islets. Transplantation 45:827–830
53. Sternberger LA, Hardy PH Jr, Cuculis JJ, Meyer HG (1970) The unlabeled antibody enzyme method of immunohistochemistry. Preparation and properties of soluble antigen-antibody complex (horseradish peroxidase – anti-horseradish peroxidase) and its use in identification in spirochetes. J Histochem Cytochem 18:315–333
54. Menger MD, Jäger S, Walter P, Hammersen F, Messmer K (1990) The microvasculature of xenogeneic transplanted islets of Langerhans. Transplant Proc 22:802–803
55. Menger MD, Sack F-U, Barker JH, Feifel G, Messmer K (1988) Quantitative analysis of microcirculatory disorders after prolonged ischemia in skeletal muscle. Therapeutic effects of prophylactic isovolemic hemodilution. Res Exp Med 188:151–165
56. Menger MD, Hammersen F, Messmer K (1990) The microcirculation of the islets of Langerhans: present state of the art. Prog Appl Microcirc 17:192–215
57. Zarem HA, Zweifach BW (1967) Development of microcirculation in full thickness autogenous skin grafts in mice. Am J Physiol 212:1081–1090
58. Demarchez M, Hartmann DJ, Prunieras M (1987) An immunohistological study of the revascularization process in human skin transplanted onto the nude mouse. Transplantation 43:896–903
59. Menger MD, Jäger S, Hammersen F, Messmer K (1991) Microvascular phenomena during pancreatic islet graft rejection. Langenbecks Arch (in press)
60. Menger MD, Jäger S, Walter P, Feifel G, Hammersen F, Messmer K (1989) Angiogenesis and hemodynamics of microvasculature of transplanted islets of Langerhans. Diabetes 38 [Suppl 1]:199–201
61. Menger MD, Jäger S, Walter P, Messmer K (1990) Influence of hyperglycemia on the process of angiogenesis and revascularization of freely transplanted islets of Langerhans. Transplant Proc 22:821–822

62. Menger MD, Jäger S, Walter P, Messmer K (1990) Influence of culture temperature (24°C/37°C) on revascularization of transplanted islets of Langerhans in the Syrian hamster. Transplant Proc 22:823
63. Menger MD, Jäger S, Walter P, Schorlemmer HU, Messmer K (1990) The influence of (±)-15-deoxyspergualin on revascularization of xenogeneic transplanted islets of Langerhans. Transplant Proc 22:1631–1632
64. Menger MD, Wolf B, Messmer K (1990) White cell endothelium interaction in islet graft rejection. Eur Surg Res 22:187–188
65. Griffith RC, Scharp DW, Hartman BK, Ballinger WF, Lacy PE (1977) A morphologic study of intrahepatic portal-vein islet isografts. Diabetes 26:201–214
66. Ledbetter DJ, Reidy MA, Trapper D (1989) Cyclosporin causes endothelial cell injury: a mechanism for vascular complications after transplantation. Surg Forum 36:268–269
67. Menger MD, Wolf B, Jäger S, Walter P, Messmer K (1990) The influence of prednisolone on revascularization of pancreatic islet grafts. Transplant Proc 22:2042–2043
68. Cotran RS (1987) New roles for the endothelium in inflammation and immunity. Am J Pathol 129:407–413

Why Should We Rise with the Lark?

J. Aschoff

In 1826 the essayist Charles Lamb criticized the British precept that "we should rise with the lark" as a popular fallacy. His advice was: "For a mere human gentleman we take ten to be the very earliest hour at which he can begin to think of abandoning his pillow" [21]. This is in strong contrast to the note which James Boswell, while travelling in Holland, made in his diary on September 25, 1763: "At present, I rise every day early, a practice which contributes much to the preservation of health because it knits up the nerves and gives hardiness and vigour to the constitution" [23]. To complicate matters, the time of going to bed could be discussed; there comes to mind the provocative assertion of Elias Canetti [15]: "A respectable man retires at nine." We may ask, however, why we have to wake up at all. And how is it that the lark gets up so early? At the time of Charles Lamb the answer would have been: the lark wakes up when, during dawn, the intensity of illumination reaches a certain, species-specific threshold – it is the alternation of light and dark which determines the rhythm of wakefulness and sleep. In the meantime we have learned that the lark does not need external signals for rising and roosting but relies upon an internal biological clock. Here, I briefly present data that indicate the existence of such a clock in animals and then discuss in more detail aspects of temporal organization in man.

To test the hypothesis of an endogenous control of the sleep-wake cycle, we recorded the hopping activity of chaffinches, kept singly in cages and exposed either to a 12-h light – 12-h dark cycle (LD 12:12) or to continuous illumination (LL). As can be seen in Fig. 1, the activity rhythm is equally well expressed in LD as in LL. The two recordings differ, however, in one important respect: while the onsets of activity are linked to light-on in LD, and hence succeed each other in 24-h intervals, they occur at increasingly earlier times of day in LL. This means that the rhythm assumes in LL a mean period of less than 24 h, which has been denoted "circadian" (from the Latin *circa*, about, and *dies*, day). The deviation of this "free-running" rhythm from the rotation of the earth around its axis justifies the conclusion that the rhythm is not controlled by external signals but is generated endogenously.

Circadian rhythms have been documented in all eukariotes from unicellular organisms to humans. They have the characteristics of self-sustaining oscillations and hence comply with the requirements of true clocks. Within certain limits of frequency they can be entrained (synchronized) by periodic factors

K. Messmer M. Stein (Eds.)
Pathways in Applied Immunology

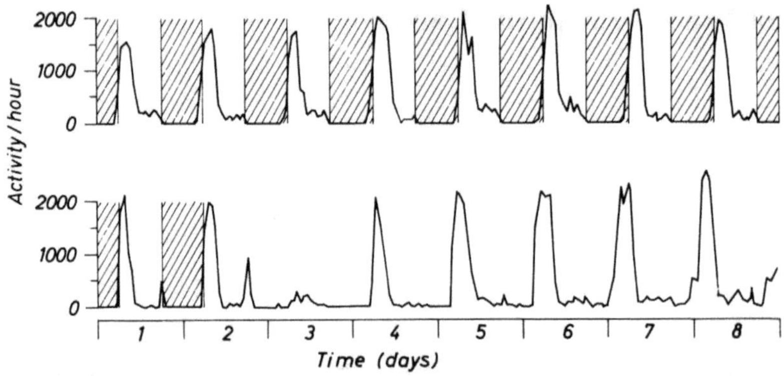

Fig. 1. Circadian activity rhythm of chaffinch kept either in a 12-h/12-h light-dark cycle (*above*) or in constant illumination (*below*). *White area*, 10 lux; *shaded area*, 0.15 lux. (Adapted from Aschoff and Pohl 1970 [9])

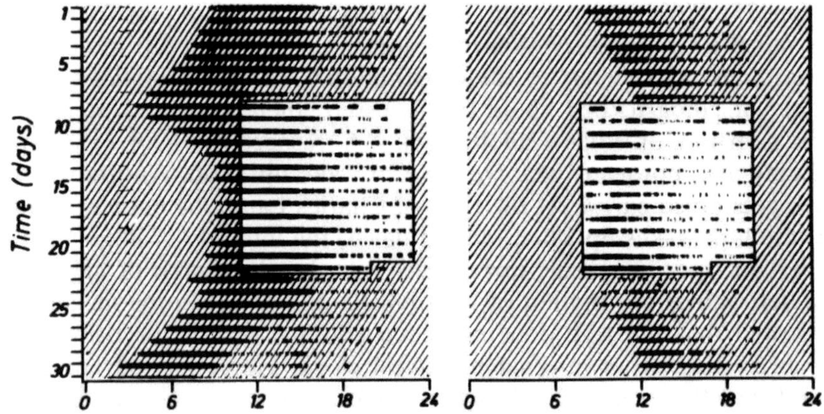

Fig. 2. Activity recordings from two chaffinches, kept alternatingly in constant dim illumination and in 12-h/12-h light-dark cycles. Strips of daily activity (*black marks*) pasted beneath each other. *White area*, 200 lux; *shaded area*, 0.5 lux. (Adapted from Aschoff and Wever 1966 [11])

in the environment, designated as zeitgebers. The most powerful zeitgeber is provided by the alternation of light and dark, but cycles of low and high ambient temperature or of social cues can entrain the rhythms of at least some organisms [1]. When entrained, the rhythm keeps a distinct phase relationship to the zeitgeber, in accordance with rules known from technical oscillators. An example is given in Fig. 2, which shows the activity rhythms of different frequencies: the period is shorter than 24 h in the left and longer than 24-h in the right actogram. After the introduction of an LD zeitgeber at day 8, both rhythms eventually become entrained, but the one with the short period keeps a leading phase to light-on, and the one with the long period a lagging phase. In other words, when entrained to 24h, the bird with a fast running clock becomes an "early riser", and the bird with a slow running clock a "late riser".

After the properties of free-running circadian rhythms and their entrainment by zeitgebers had been described in animals, the question suggested itself whether these findings were applicable to man. To study this we needed a room where subjects could be shielded from all possible zeitgebers, including noise from people, traffic, and birds. Our problems were solved when Walter Brendel pointed out to us that at the clinic in Munich there were underground facilities which had been used during the war for emergency surgery. They were designated for his group to move into later and could in the meantime easily be accommodated to our purposes. It is thus due to Walter Brendel that we got a chance to explore the feasability of long-term isolation studies. The positive results of the first ten experiments encouraged us to construct a specifically designed isolation unit next to the institute at Andechs. It consists of a reasonably sized living room with bed, a small kitchen and a toilet with shower. There our subjects live for several weeks, singly, without any information on time of day, either in LL or by turning the lights in the unit on and off as they like. By pressing buttons, they indicate the times when they wake up, when they take a meal (prepared by themselves), and when they go to sleep (turning off their bedside reading lamp). Body temperature is recorded continuously by means of a rectal probe.

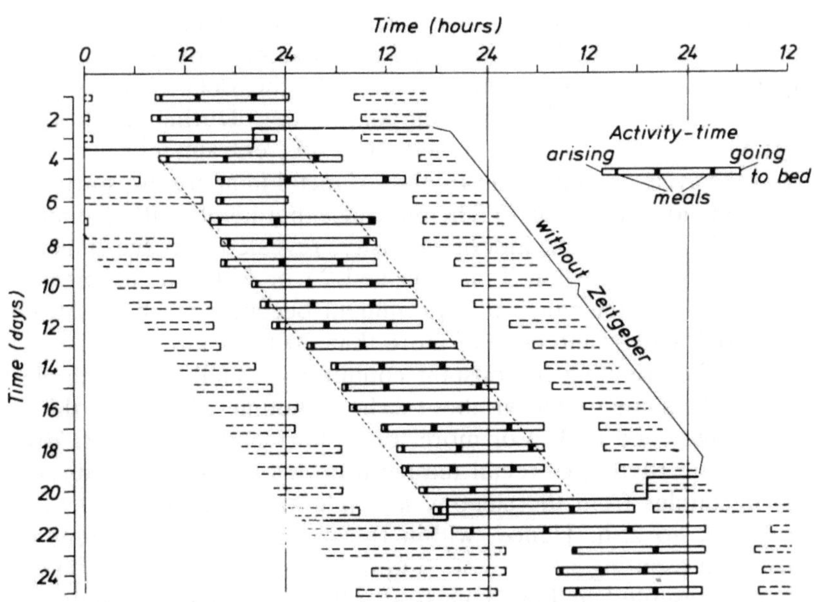

Fig. 3. Free-running circadian rhythm of a human subject in isolation. Successive times of wakefulness (*white bars*) plotted beneath each other as a function of time of day. Day 1: August 5, 1961. (From Aschoff and Wever 1962 [10])

The results of one of the first experiments, still being performed in Munich, are presented in Fig. 3. During the initial 3 days the subject knew the time of day and was hence synchronized to 24 h. After all temporal cues had been removed in the evening of day 3, successive awakenings occurred at increasingly later times of day, resulting in a mean circadian period of 25.9 h, with 17.5 h of wakefulness and 8.4 h of sleep. When isolation was terminated at day 22, the subject readjusted his rhythm to local time by a large delay shift. The recording shown in Fig. 3 is representative of most of the experiments in that during isolation, the frequency of the rhythm slows down. In a sample of 147 subjects, the overall mean of the period was found to be 25.0 h, with an interindividual standard deviation of only ±0.5 h [27]. From the tendency of the human clock to be slow it results that daily advance corrections are needed to achieve synchronization with the 24-h day. The "lagging behind" of the rhythm explains why the jet-lag symptoms are usually less severe after a westbound than after an eastbound flight, and why late risers prevail over early risers.

In the isolation unit, human circadian rhythms can be entrained by means of appropriate zeitgebes such as LD cycles complemented by regular gong signals which provide an additional time structure [25]. The results presented in Fig. 4 are from an experiment in which the subject could make use of small reading lamps at will and hence retained some independence of the main lighting conditions. In the initial 24-h day, the subject woke up several hours after lights-on, and retired at about midnight. After the period T of the zeitgeber had been lengthened to 26.7 h, he changed into an early riser, eventually waking up (to his own surprise, and not knowing why) before the lights had been turned on. Concurrently the maxima and minima of rectal temperature were also advanced relative to the sleep-wake cycle, indicating drastic changes in the internal temporal order. During the last part of the experiment, when T was shortened to 22.7 h, entrainment was lost, and the sleep-wake cycle as well as the rhythm of rectal temperature started to free-run with a common period of 25.2 h. Further experiments with similar zeitgebers, i.e. with light of normal indoor intensity, revealed that the limits of entrainment were reached when T approached 23 or 27 h. In the meantime it has been shown that the range of entrainment becomes much larger when bright light is used with an intensity of more than 2000 lux [28].

When free-running in constant conditions, usually all rhythmic functions adhere to the same period of about 25 h. Sometimes, however, the sleep-wake cycle is lengthened to 30 h or more, or is shortened to less than 20 h. In these instances, the rhythm of rectal temperature (as that of other autonomic functions) uncouples from the sleep-wake cycle and continues to free-run with a period close to 25 h. Hence, within the organism, two different frequencies coexist, representing a state of "internal desynchronization" [12]. The two diagrams presented in Fig. 5 illustrate "long desynchronization" (above) by a lengthening, and "short desynchronization" (below) by a shortening of the sleep-wake cycle. As yet, internal desynchronization has been documented only in subjects living in isolation, but one wonders whether it may also occur

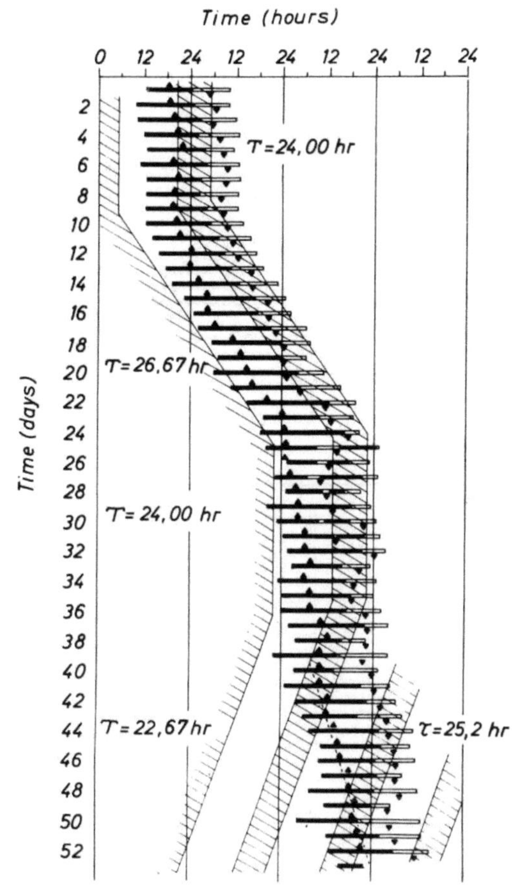

Fig. 4. Circadian rhythms of wakefulness (*black bars*) and sleep (*white bars*) and of rectal temperature (*triangles above and below bars* for maxima and minima, respectively). Data from a subject who was exposed, in the isolation unit, to an artificial light-dark cycle complemented by gong signals; reading lamps were at the subject's disposal. *Shaded area*, darkness: *T*, zeitgeber period; τ, circadian period. (From Aschoff et al., 1969 [14])

under normal living conditions, for example, if a person lacks the proper responsiveness to the zeitgebers. On January 16, 1922, Franz Kafka [20] made the observation: "The clocks don't agree. The inner one rushes along in a devilish way, while the outer one goes its accustomed pace" (my translation from German).

The occurrence of internal desynchronization has given rise to the hypothesis that the human circadian system consists of at least two oscillators, a "weak" one controlling the sleep-wake cycle, and a "strong" one controlling autonomic functions [27]. The two oscillators interact and are usually coupled to each other. During desynchronization, they still influence each other mutually. This is shown by regular modulations of the amplitude in the rhythm of rectal temperature [12] as well as the duration of sleep [31], depending on

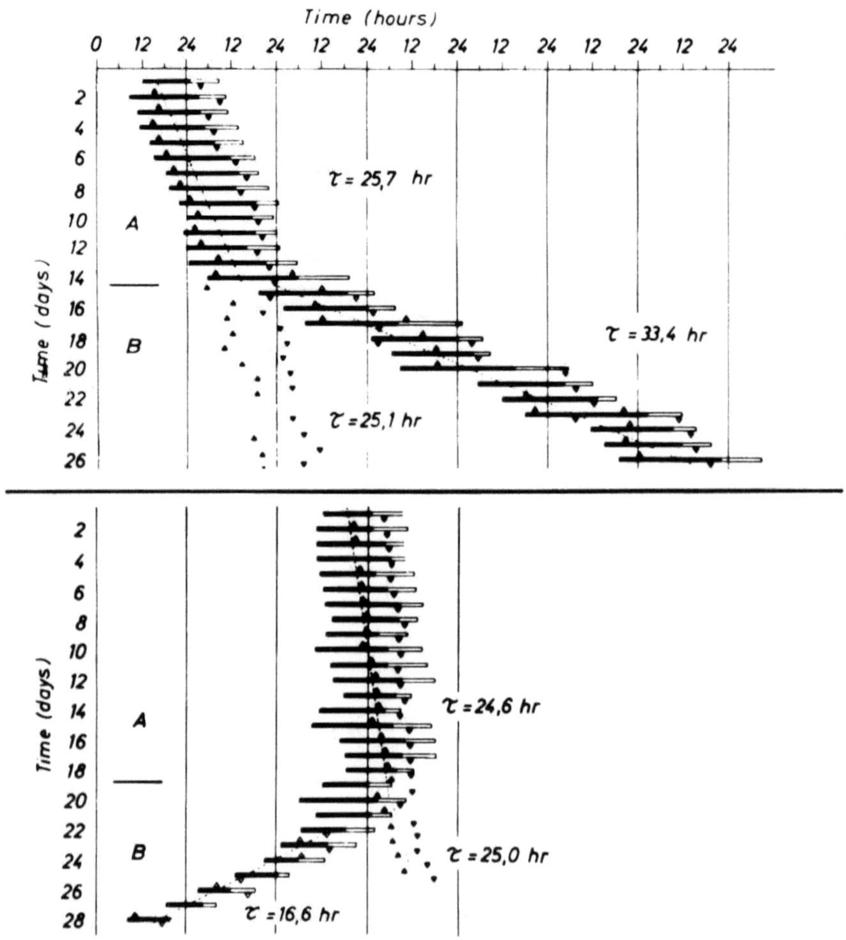

Fig. 5. Circadian rhythms of wakefulness (*black bars*) and sleep (*white bars*) and of rectal temperature (*triangles* for maxima and minima). Data from two subjects who lived singly in the isolation unit in constant conditions. *Part A*, internal synchronization; *part B*, internal desynchronization (*above*, long desynchronization; *below*, short desynchronization). τ, Circadian period. (From Wever 1975 [26])

whether the rhythms are in phase or out of phase with each other. Various two-oscillator models have been developed to describe these features [22]. It must be noted, however, that a more recent model which postulates a recovery process in combination with clock-controlled thresholds for onset and end of sleep also seems able to simulate all the phenomena mentioned [16].

During internal desynchronization a subject may stay awake continuously for up to 40 h (long desynchronization) or for only 12 h (short desynchronization). Surprisingly, the subjects are almost never aware of their unusual way of living, but express at the end of the experiment their belief to have followed a more or less regular 24-h schedule of wakefulness and sleep. This

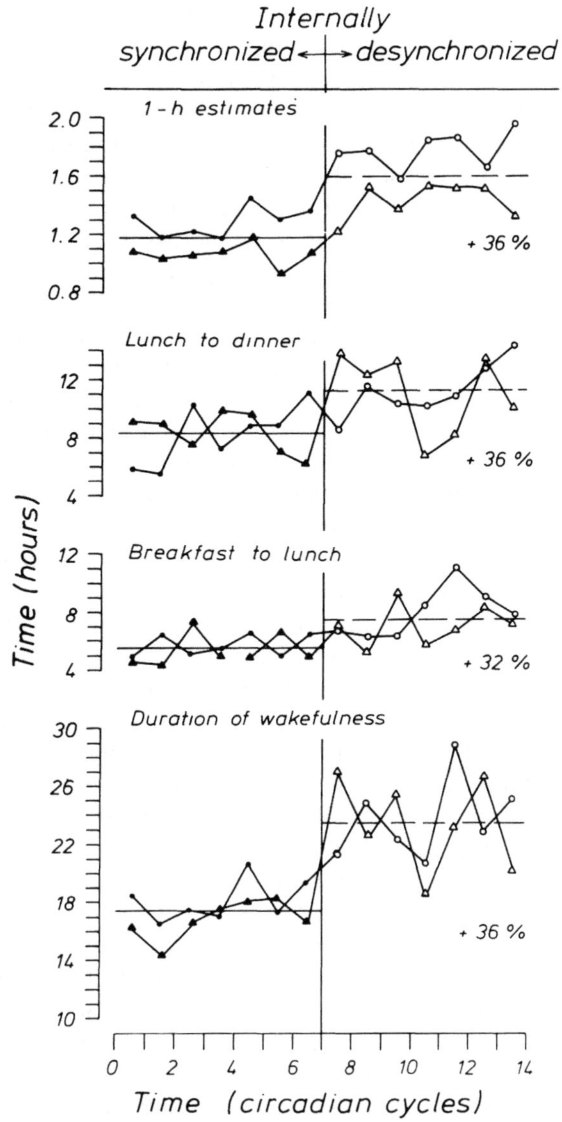

Fig. 6. Duration of wakefulness, intervals from breakfast to lunch and from lunch to dinner, and 1-h estimates recorded during 14 consecutive circadian cycles in two subjects who, during isolation, changed from internal synchronization to long desynchronization. *Numbers at curves*, increases in percentage of the mean, averaged from both subjects (*dotted lines*). (From Aschoff 1986 [6])

is due to a drastic change in their perception of time. In a series of experiments we asked the subjects to press a button whenever they thought that 1 h had elapsed. The intervals produced in this manner were almost proportional to the duration of wakefulness. Consequently, the subjects experienced about the same number of "hours" on short as on long "days". There were large interindividual differences in the means of 1-h estimates, ranging from 1.1 to 2.06 h [4]. These findings support the notion of a "personal speed" in the passage of time, as expressed by William Shakespeare in *As You Like It*: "Time travels in divers paces with divers persons."

The large variations in the duration of wakefulness raise the further question of how the subjects organize their meals on long and short "days". Under normal living conditions, we usually become hungry again 4–5 h after having eaten last. Hence, it could be assumed that a subject who is awake for 32 h instead of the normal 16 h ingests three extra meals, and that a subject who is awake for only 12 h skips one meal. This did not happen in our experiments. Instead of altering the number of meals, the subjects adjusted the intervals between meals to the duration of wakefulness α. This is shown in Fig. 6 by the data from two subjects who, after seven cycles of internal synchronization, changed to long desynchronization. The increase in α by 36% is accompanied by an equal lengthening of the intervals from breakfast to lunch and from lunch to dinner, and of the daily means in 1-h estimates. In a further step of the analysis, we computed for each subject the means of inter-meal intervals and of 1-h estimates, averaging separately all cycles with internal synchronization, long desynchronization, and short desynchronization. The results of such computation, summarized in Fig. 7, demonstrate a strong positive correlation of the intervals and of the 1-h estimates with α. Somehow, this dependence on α is least expected for the interval from waking up to breakfast because one might assume that breakfast is taken within a relatively short "standardized" time after awakening, according to personal habits. But even this first interval of a day's routine is closely related to α. It seems as if the organism "knows" already at the time of waking, for how long wakefulness will last that particular day.

If, during internal desynchronization, the inter-meal intervals become twice as long or two-thirds as long as they are during ordinary life, one should expect changes either in the amount of food eaten or in body weight. Neither of these assumptions has been confirmed. Body weight remains stable [7], and the caloric intake per meal was the same in internally synchronized and in desynchronized subjects [17]. From these findings it must be concluded that energy expenditure is adjusted to α (or to the length of the cycle). As we have discovered recently, to some extent this is achieved by a negative correlation between the hourly locomotor activity and α. Due to this "homeostatic control", the total amount of activity per "day" remains constant irrespective of large variations in α [8]. Since under the living conditions provided in the isolation unit only about one-third of the total energy expenditure can be attributed to muscular work, additional energy must be conserved on long "days" by other means. It remains to be seen whether basal metabolic rate is negatively correlated with α [5], and at present nothing can be said about the underlying control mechanisms. Even less is known about the mechanisms which relate time perception to α. Undoubtedly, Virginia Woolf [30] is right in her observation that the "discrepancy between time on the clock and time in the mind is less known than it should be and deserves fuller investigation."

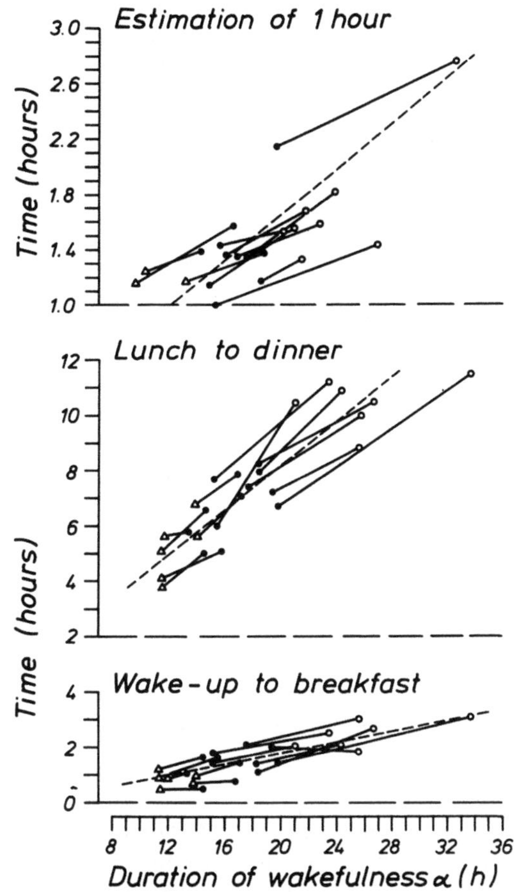

Fig. 7. Means of the intervals from waking up to breakfast and from lunch to dinner, and of 1-h estimates, drawn as a function of the duration of wakefulness. Data averaged from all cycles of internally synchronized rhythms (*closed circles*), of long desynchronized rhythms (*open circles*), and of short desynchronized rhythms (*open triangles*). Lines connect data from the same subject. *Oblique dashed lines* represent proportionality with the duration of wake time. (After Aschoff 1985a [4] and Aschoff et al., 1986 [13])

In returning to the initial problem of this presentation, we may ultimately ask the question: why is it so difficult to get up in the morning; the German psychiatrist Hoche [19] noted "Waking up is the most difficult moment of the day" (my translation from German). And furthermore, why is it that waking up becomes even more difficult in winter than in summer? Can it be that our need for sleep depends on the season? Seasonal rhythms in the physiology and psychology of man are well documented [2]. The few examples provided in Fig. 8 demonstrate that, contrary to expectation, rates of suicides reach a maximum early in summer, at about the time when most children are conceived. Mirror images of these curves are those representing the circulation of books taken out of the central libraries of three major cities. It is up to

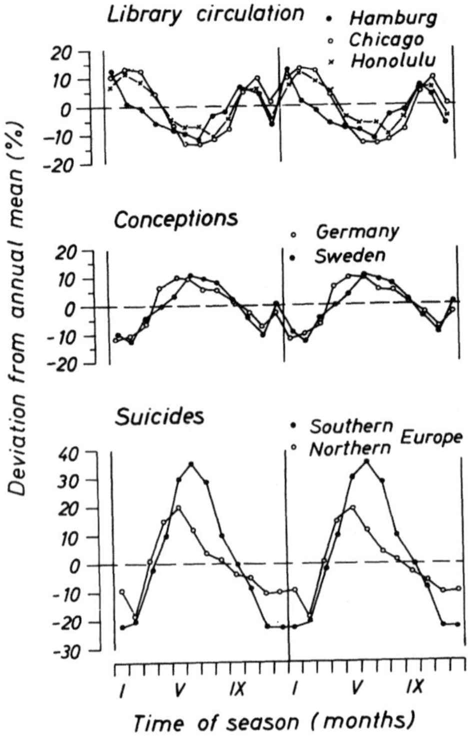

Fig. 8. Seasonal rhythms in the monthly rates of suicides, conceptions, and library circulations. Means from several years, plotted twice along the abscissa. (From Aschoff 1981c [3])

the reader to draw any conclusion he likes from the negative correlation between reading and love making. A seasonal rhythm in sleep propensity has not been described so far, and it would be difficult to document it in everyday life where social constrains limit our sleep. Over against this, in our isolation unit the subjects were unrestrained in the timing of their sleep. Since experiments were made all year round, we could study the effects of seasons on the duration of sleep [29]. We extracted from our recordings the fraction of sleep in each circadian cycle and computed 2-monthly means separately for women and for men. The results (Fig. 9) demonstrate a seasonal modulation in the sleep-wake cycle, with an increase in the duration of sleep from a summer minimum to a winter maximum. Apparently, humans have from their mammalian ancestors the remains not only of a seasonal rhythm in reproduction [24] but also of a tendency to hibernate. Interesting as these data are with regard to the possibility of a "circannual" clock in humans, they carry a further information. As can be seen in Fig. 9, women sleep longer than men at all times of the year. This observation is of interest in view of the fact that there seems to be a positive correlation between the habitual duration of sleep and intelligence [18]. Husbands, though less intelligent, should be sensible enough to abandon their pillow softly and to let their wives sleep as long as they like.

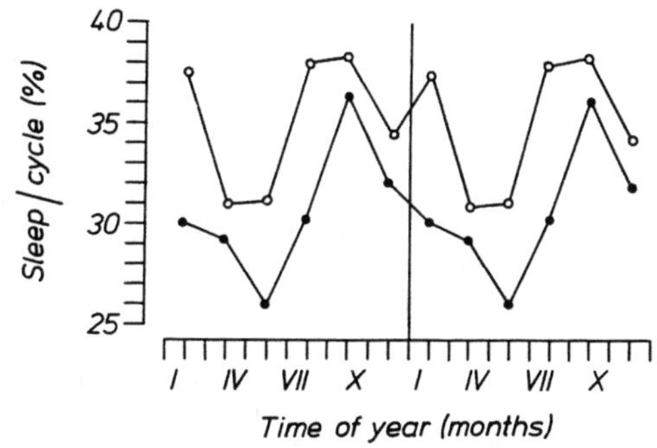

Fig. 9. Seasonal variations in the fraction of sleep per circadian cycle, recorded in the isolation unit in the free-running rhythms of women (*open circles*) and men (*closed circles*). Each point represents the average of 2-monthly means from 6–12 subjects. (Data from Wirz-Justice et al., 1984 [29])

References

1. Aschoff J (1981) Freerunning and entrained circadian rhythms. In: Aschoff J (ed) Biological rhythms. Plenum, New York, pp 81–93 (Handbook of behavioral neurobiology, Vol 4)
2. Aschoff J (1981) Annual rhythms in man. In: Aschoff J (ed) Biological rhythms. Plenum, New York, pp 475–487 (Handbook of behavioral neurobiology, Vol 4)
3. Aschoff J (1981) Twenty years on. In: Follett BK, Follett DE (eds) Biological clocks in seasonal reproductive cycles. Wright, Bristol, pp 277–288
4. Aschoff J (1985) On the perception of time during prolonged temporal isolation. Hum Neurobiol 4:41–52
5. Aschoff J (1985) Time perception and the timing of meals during temporal isolation. In: Hiroshige T, Honma K (eds) Circadian clocks and zeitgebers. Hokkaido University Press, Sapporo, pp 3–18
6. Aschoff J (1986) Circadiane Systeme. Physiol Aktuel 2:5–19
7. Aschoff J (1989) Temporal orientation: circadian clocks in animals and humans. Animal Behav 37:881–896
8. Aschoff J (1990) Interdependence between locomotor activity and duration of wakefulness in humans during isolation. Experientia 46:870–871
9. Aschoff J, Pohl H (1970) Rhythmic variations in energy metabolism. Fed Proc 29: 1541–1552
10. Aschoff J, Wever R (1962) Spontanperiodik des Menschen bei Ausschluss aller Zeitgeber. Naturwissenschaften 49:337–342
11. Aschoff J, Wever R (1966) Circadian period and phase-angle difference in chaffinches (*Fringilla coelebs* L.). Comp Biochem Physiol 18:397–404
12. Aschoff J, Gerecke U, Wever R (1967) Desynchronization of human circadian rhythms. Jpn J Physiol 17:450–457
13. Aschoff J, von Goetz C, Wildgruber C, Wever RA (1986) Meal timing in man during isolation without time cues. J Biol Rhythms 1:154–162
14. Aschoff J, Pöppel E, Wever R (1969) Circadiane Periodik des Menschen unter dem Einfluss von Licht-Dunkel-Wechseln unterschiedlicher Periode. Pflügers Arch 306:58–70
15. Canetti E (1935) Die Blendung. Fischer, Frankfurt
16. Daan S, Beersma DGM, Borbely AA (1984) Timing of human sleep: recovery process gated by a circadian pacemaker. Am J Physiol 246:R161–R178

17. Green J, Pollack CP, Smith GP (1987) Meal size and intermeal interval in human subjects in time isolation. Physiol Behav 41:141–147
18. Hicks RA, Pellegrini RJ, Cavanaugh A, Sahatjian J, Sandham L (1978) Fluid intelligence levels of short- and long-sleeping college students. Psychol Rep 43:1325–1326
19. Hoche AE (1936) Einer Liebe Weg. Reissner, Dresden
20. Kafka F (1973) Tagebücher 1910–1923. Fischer, Frankfurt
21. Lucas EV (ed) (1903) The works of Charles Mary Lamb, Vol 2: Elia and the last essays of Elia. Methuen, London. Reprinted 1968 by Ames, New York
22. Moore-Ede M, Czeisler CA (eds) (1984) Mathematical models of the circadian sleep-wake cycle. Raven, New York
23. Pottle FA (ed) (1952) Boswell in Holland 1763–1764. Heinemann, London
24. Roenneberg T, Aschoff J (1990) Annual rhythm of human reproduction. J Biol Rhythms 5:195–216 and 217–239
25. Wever R (1970) Zur Zeitgeber-Stärke eines Licht-Dunkel-Wechsels für die circadiane Periodik des Menschen. Pflügers Arch 321:133–142
26. Wever R (1975) The circadian multi-oscillator system of man. Int J Chronobiol 3:19–55
27. Wever R (1979) The circadian system of man. Springer, Berlin Heidelberg New York
28. Wever R, Polasek J, Wildgruber CM (1983) Bright light affects human circadian rhythms. Pflügers Arch 396:85–87
29. Wirz-Justice A, Wever R, Aschoff J (1984) Seasonality in freerunning circadian rhythms in man. Naturwissenschaften 71:316–319
30. Woolf V (1946) Orlando. Penguin, New York
31. Zulley J, Wever R, Aschoff J (1981) The dependence of onset and duration of sleep on the circadian rhythm of rectal temperature. Pflügers Arch 391:314–318

Subject Index

MIX
Papier aus verantwortungsvollen Quellen
Paper from responsible sources
FSC® C105338

If you have any concerns about our products,
you can contact us on
ProductSafety@springernature.com

In case Publisher is established outside the EU,
the EU authorized representative is:
Springer Nature Customer Service Center GmbH
Europaplatz 3, 69115 Heidelberg, Germany

Printed by Libri Plureos GmbH
in Hamburg, Germany